Advanced Topics in Science and Technology in China

Zhejiang University is one of the leading universities in China. In Advanced Topics in Science and Technology in China, Zhejiang University Press and Springer jointly publish monographs by Chinese scholars and professors, as well as invited authors and editors from abroad who are outstanding experts and scholars in their fields. This series will be of interest to researchers, lecturers, and graduate students alike.

Advanced Topics in Science and Technology in China aims to present the latest and most cutting-edge theories, techniques, and methodologies in various research areas in China. It covers all disciplines in the fields of natural science and technology, including but not limited to, computer science, materials science, life sciences, engineering, environmental sciences, mathematics, and physics.

More information about this series at http://www.springer.com/series/7887

Changming Du · Rongliang Qiu
Jujun Ruan

Plasma Fluidized Bed

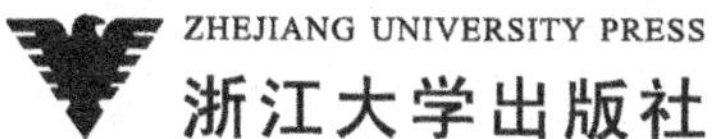

Changming Du
School of Environmental Science
  and Engineering
Sun Yat-sen University
Guangzhou, Guangdong
China

Jujun Ruan
School of Environmental Science
  and Engineering
Sun Yat-sen University
Guangzhou, Guangdong
China

Rongliang Qiu
School of Environmental Science
  and Engineering
Sun Yat-sen University
Guangzhou, Guangdong
China

ISSN 1995-6819   ISSN 1995-6827   (electronic)
Advanced Topics in Science and Technology in China
ISBN 978-981-13-3862-5   ISBN 978-981-10-5819-6   (eBook)
https://doi.org/10.1007/978-981-10-5819-6

Jointly published with Zhejiang University Press

The print edition is not for sale in China Mainland. Customers from China Mainland please order the print book from: Zhejiang University Press

Preface

Over several decades, industrial development and scientific research areas have seen the rapid extension of plasma fluidized bed in different areas, all of which are very fundamental to the industrious development in twenty-first century.

Plasma fluidized bed is an innovative tool and generally combines plasma process with an efficient reactor, fluidized bed, providing an excellent method for particulate processes over conventional technology. Plasma fluidized bed reactors for gas–solid reactions have several advantages such as enhanced heat and mass transfer, generation of extremely high temperature using plasma as a heat source or generation of chemically active species under a mild condition.

Proposal of the conception of plasma fluidized bed can be dated back to the 1960s, and a DC (direct current) torch plasma fluidized bed was firstly developed for the quenching of plasma reaction and then electrothermal plasma fluidized bed. In the 1980s and 1990s till the twenty-first century, various plasma fluidized beds, such as corona plasma fluidized bed, microwave (MW) plasma fluidized bed, radio frequency (RF) plasma fluidized bed, gliding arc (GD) plasma fluidized bed, and dielectric barriers discharge (DBD) plasma fluidized bed, have arisen gradually, and meanwhile, researches on the study and the application of plasma fluidized bed developed rapidly, which can be presented in material, metallurgy, energy, and environment fields.

The previous studies presented that plasma fluidized beds are promising in a broad industrial field, such as material processing, energy industry, environmental protection, and metallurgy industry. However, the association of the main advantages of both fluidized bed and plasma has been an exciting challenge for the further development of the reactor. And all these developments are now limited by the relative scarce of knowledge of basic properties of plasma fluidized bed and by some pending technical problems due to the limitations of the reactors. Therefore, a survey over plasma fluidized bed might be in need. In the present review, the contribution focuses on a detailed overview of previous plasma fluidized bed, including the designs of various types of plasma fluidized bed and the factors influencing overall performance. Also, the characteristics of plasma fluidized bed, i.e., the discharge characteristics, the hydrodynamic and mixing behavior of particles, and the heat transfer in the plasma fluidized bed are discussed for further understanding of the plasma fluidized beds. Finally, several research directions for the further development of the application of plasma fluidized bed reactor were proposed.

Different plasma reactors have been developed for different applications in our laboratory, including the treatment of wastewater polluted by microorganism, organics as well as heavy metal, the abatement of the organic exhausted gas, reformation of various fuel for syngas. Recently, our investigations are focused on a special type of plasma fluidized bed for plasma catalytic process as well as the organic-contaminated soil remediation, which has arisen the great interest of some researchers, and it is found that plasma fluidized bed is of great potential for some particulate process; therefore, we make an endeavor to give an overall overview of plasma fluidized bed.

All research works were supported by the Guangdong Applied Science and Technology Research Project (2015B020237005), the GuangDong Public Welfare Research and Capacity Building Project (2015A020215013), the Natural Science Foundation of GuangDong (2016A030313221), and the National Natural Science Foundation of China (50908237). The authors gratefully acknowledge the work of DanYan Ma and HanTing Huang during their research.

Guangzhou, China   Changming Du

Contents

Chapter 1
Plasma and Plasma Fluidized Bed

Abstract In this chapter, the topic of plasma is introduced and a comprehensive description of plasma fluidized bed is given. The development and application of thermal plasma and non-thermal plasma are briefly introduced. Then, the development course of plasma fluidized bed is introduced in chronological order. What's more, the principle of the removal of pollutants by plasma fluidized bed is described, and the advantages of the plasma fluidized bed are listed in detail. It's proved that the plasma fluidized bed has great potential. In the end, three sets of parameters of plasma fluidized bed are evaluated for the use of plasma fluidized bed.

Keywords Principle · Application · Fluidized bed

1.1 Plasma

Plasma, which has gained widespread application in a rater board field, is a partially or fully ionized gas consisting of various particles, such as electrons, ions, atoms, and molecules, which are electrical neutral from a macroscopic point of view. The degree of ionization ranges from low, like low temperature discharge, to very high, such as in fusion experiments. In these different cases, plasma gas shows different physical and chemical properties in different conditions. Therefore, plasma is known as the fourth state of matter, beside the solid, fluid and gas state. Different plasma systems can be categorized based on the electron number density n_e and the electron temperature T_e into two major categories, namely thermal and non-thermal plasma (Nehra et al. 2008; Attri et al. 2013).

1.1.1 Thermal Plasma

Thermal plasma (usually arc discharges, torches or radio frequency and microwave plasma) is associated with sufficient energy introduced to allow plasma constituent

© Springer Nature Singapore Pte Ltd. and Zhejiang University Press 2018
C. Du et al., *Plasma Fluidized Bed*, Advanced Topics in Science and Technology in China, https://doi.org/10.1007/978-981-10-5819-6_1

to be in thermal equilibrium (Heberlein and Murphy 2008). Generally, thermal plasma should be sustained by introducing high electrical energy, therefore the macro temperature of the thermal plasma is very high, to illustrate, a plasma torch is able to generate temperature varying from 7000 to 10,000 K (Pfender 1999), thermal plasma can be characterized by high temperature, high energy density and high reactivity so that it has been successfully developed and utilized for a board industry fields, especially for high temperature process, such as extractive metallurgy, ceramic and glass application, energy industry, chemical production process and environmental protection areas, which have shown unique advantages and great application potential, as shown in Table 1.1 (Pfender 1999; Nezu et al. 2003; Matsumoto et al. 1987; Taylor and Pirzada 1994).

Conventional plasma reactors for high temperature operations include electrothermal smelting method or plasma furnace and usually induce some problems. On one hand, in order to keep the reaction zone in high temperature, energy density demanded to maintain the plasma discharge would be very large, also, the heat losses are very serious, and therefore the utilization of input energy is very low (Pfender 1999). On the other hand, the transportation process of heat and mass between injected plasma gas and the solid materials are relatively low, and as a result the kinetic process requires a very long resident time (Pfender 1999). The overheating of the local position which would lead to the instability or even the damage of the reactor may be another problem. Hence, the high-temperature operation in conventional plasma reactor demands a high capital and much more energy input. To solve the problems for scaling up of the application of plasma process, enhanced heat and mass transfer are problems to deal with.

Table 1.1 Typical application of thermal plasma

Application field	Process	Advantages
Metallurgical process	Production of alloys and refractory materials	Controllable, stable, low loss of alloy content, suitable for the production of the metal hard to melt, easy operation, low cost and less pollution
Energy industry	Plasma gasification Energy reformation	High efficiency, small size of the set-up, easy operation and environment friendly
Chemical process	Synthesis of chemicals such as acetylene, NO_x and phosphates and ultra fine powers	Uniform size, high purity, enriched species, easy operation, new products
Waste treatment	Treatment of unclear waste, solid waste and toxic matter	Low cost and easy operation, high detoxification rate
Waste recycling	Recovery of catalyst, Recovery of metals from iron and steel industries	High recycle rate, less energy consumption, easy operation

1.1.2 Non-thermal Plasma

Much different from thermal plasma, non-thermal plasma is obtained using lower power (usually corona discharge, dielectric barrier discharge, gliding arc discharge, glow discharge and spark discharge), which features an energetic electron temperature much higher than that of the bulk gas molecules. Within non-thermal plasma, energetic electrons can collide with background molecules (N_2, O_2, H_2O, etc.) to generate secondary electrons, photons, ions and radicals. Therefore, in non-thermal plasma system, a highly active chemistry atmosphere is generated featured by high selectivity and energy efficiency in plasma chemical reactions. Non-thermal plasma has attracted more and more attention for its significant advantages over other conventional technologies. This is because non-thermal plasma is far from equilibrium which results in an environment with high concentration of chemically active species and gas temperature staying relatively low or even at room temperature. This feature gives birth to a totally new chemical environmental for chemical process. As a result, non-thermal plasma infers a relatively active chemical atmosphere under lower temperature, such as clean energy production, food engineering, pollution abatement, chemical synthesis, surface modification and nanoscience (Yamamoto 1997; Christodoulatos et al. 2004; Allman et al. 1987; Laroussi 2009; Şen et al. 2012).

Similar to thermal plasma process, for particulate processing within non-thermal plasma reactor, it is important to achieve an intimate mixing of the powders with the gas phase since heat and mass transfer coefficients scale with the exposed particle surface area to the gas. However, many plasma reactors, drum or batch-type, are often faced with the following drawbacks due to the lack of solid mixing with the flowing gas: long operation periods and unsatisfactory broad particle residence time distribution, weak particle-gas contact as well as non-uniform operation. Also, agglomeration of fine powder particles (below 100 μm), and discontinuous (batch-wise) processing are both problems to be dealt with for the further development of non-thermal plasma in respective to particulate operation.

Therefore, for better utilization of plasma reactor for gas-solid reaction, an appropriate gas-solid reactor system has to be applied for the best dispersion of the solid in the plasma gas, which is similar to the design of criteria of fluidized bed.

1.2 Fluidized Bed

Gas-solid processes are important in most industrial applications, examples being energy conversion, environmental protection, food manufacturing. The efficient interaction between the solid particles and gas flow should be a major objective of gas-solid processing, which can be realized by fluidization. Fluidization means injecting a fluid into a bed of particles, which enable the bed of particles move with the injected fluid in a fluid state. For specific gas-solid particles, the state of

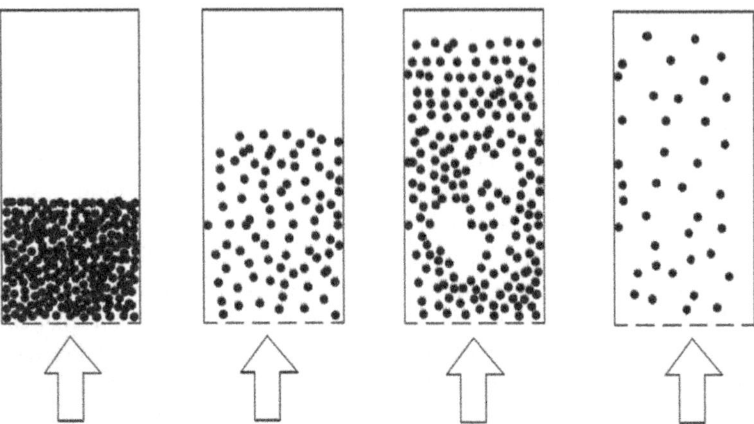

Fig. 1.1 The flow rate of the fluidized bed is changed with the flow velocity

fluidization depends on the gas velocity. As the injected fluid velocity increased from to a larger value, the state of the bed of particles varies a lot, which can be seen in Fig. 1.1 (Renzo and Maio 2007), when the fluid velocity is very small, the pressure across the bed is also relatively low and cannot balance the weight of the bed of particles, the bed is in still state. With the velocity increasing, bubbles appear in the bed. With the gas velocity further increasing, the pressure also increases until it balances the weight of the bed, and all particles are suspended in the upward gas (Thorley et al. 1959). This is defined as minimum fluidization, and the gas velocity at this point is called minimum fluidization velocity. Previous papers presented that for the gas-solid process, a larger gas velocity beyond minimum fluidization makes the fluidization become more violent and leads to instabilities with bubbling and channeling of the gas. Researchers defined this type of reactor as an aggregative fluidized bed, a heterogeneously fluidized bed, a bubbling fluidized bed or simply a gas fluidized bed (Renzo and Maio 2007).

Fluidized bed has opened its way over the industrial application since 1922, when Winkler developed the first fluidized bed for coal gasification. Based on the injected gas velocity chosen in operation, fluidized bed can be classified as fixed bed, bubbling bed, turbulent bed, fast bed and gas pneumatic transport, which are shown in Fig. 1.2. In the fluidized bed, particles can be mixed well with the gas contact intensively with the injected gas. In a fluidized bed, solid particles can act like fluid, which can significantly improve some physical and chemical features of the solid particles. Such a gas–solid system improves the contact surface and the contact efficiency between gas flow and injected particles significantly. Therefore, fluidized bed systems are characterized by a perfect mixing of particles, a high rate of heat and mass transfer and a continuity of solid processing. Within a fluidized bed, gas-solid reaction can be distributed in the overall working space. It is thus not surprising that fluidized bed reactors are widely used in the chemical and metallurgical industries for transportation and enhanced gas-solid interaction.

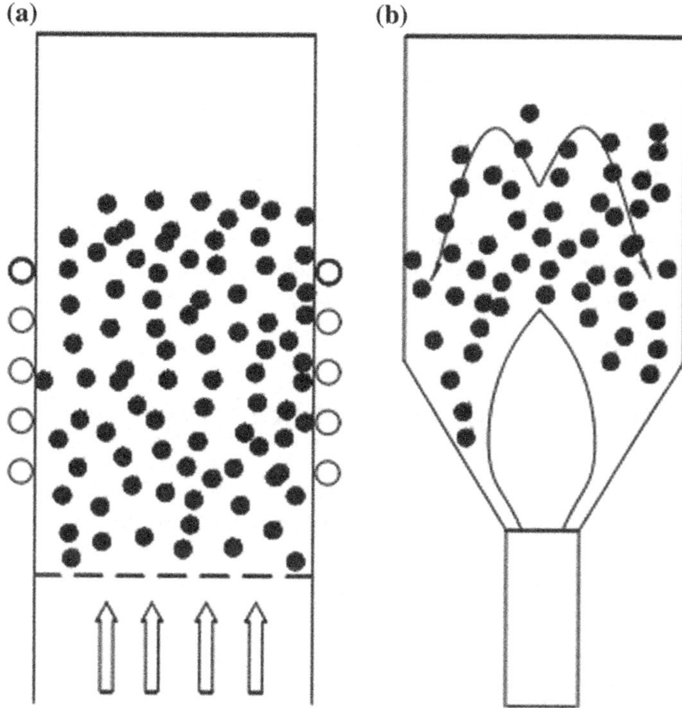

Fig. 1.2 **a** plasma fluidized bed and **b** the typical structure of plasma jet bed

Spouted bed, a special fluidized bed with spouting particles, was initially designed in 1954 by Mathur and Gishler at the National Research Council of Canada for the drying of wheat (Mathur and Gishler 1955). Since then, numerous other applications have been found (Thorley et al. 1959). Compared with conventional fluidized bed, spouted bed deals with larger particles and the injected gas is usually a jet with a large velocity. A typical spouted bed usually comprises a central gas entraining solids in an upward motion and a jet, slow counter current flow in of solids in the surrounding annulus. Entrained particles in the jet rise above the bed level and fall back out. The annular region is between the spout and bed wall. The particle then slowly travels downward and radically inward towards the spout in the lower part of the annulus. Solid particles are circulated through the bed resulting efficient solid-fluid contact and good solid mixing. Flat bottomed beds have been designed; particles in spouted beds are typically cone shaped to permit to fat slowly towards the spout. Otherwise, a dead zone is formed in the bottom region where particles may accumulate (Ma et al. 2013).

In conventional gas-solids fluidized bed operations and spouted beds, particles are suspended by up-flowing gas streams against the force of gravity, which results in many advantages to the processes, e.g., enhanced mass and heat transfer and improved inter-phase contact, but also leads to heterogeneous flow structure and significant back mixing of phases. However, there also existed various special fluidized bed for specific gas-solid reaction, to illustrate, downer bed has been proposed and investigated universally to achieve highly selective catalytic conversions and to avoid unwanted by-products in the industries, a reactor allowing for very short contact time between phases and precisely controlled gas and solid catalyst residence times, while in the field of particle processing the circulating fluidized bed represents an efficient reactor concept (Arpagaus et al. 2005). The law density and the homogeneous dispersion of the particulate phase in the gas phase, as well as the high relative velocity provide an intense mass and heat transfer between the gas and the solid phase, which allows high reaction rates.

1.3 Principle of Plasma Fluidized Bed

As mentioned above, plasma chemistry and fluidized bed technology are among the most advanced methods of material treatment and other industrial applications with respect to the chemical reaction and physical transportation. An idea to integrate the plasma discharge and fluidized bed into one system seems to an excellent approach to solve the mixing and heat transfer problem of plasma process for gas-solid processing. Fluidized bed reactors were initially brought into plasma application mainly for the quenching of the plasma by contacting the fluidized solids with the tail end of the DC or RF plasma jet or just to form a spouted bed by the plasma gas jets (Goldberger and Oxley 1963). Since then, investigations upon plasma fluidized bed have grown rapidly. Typical configurations of plasma fluidized/spouted bed can be described in Fig. 1.2a and b respectively (Flamant 1994).

In a plasma fluidized bed reactor, fluidized solid particles contact with the active and current-carrying portion of the plasma solid reactants. For a plasma fluidized bed, electrodes are usually located along the reactor chamber, connected with high voltage power supply to induce the discharge inside the reactor. Fluidization gas and plasma gas can be introduced through the distributor at the bottom of the fluidized bed, and mostly plasma gas was also used as fluidization to simplify the structure. While for plasma spouted bed, mostly plasma torch or gliding arc plasma was used as plasma source.

The fluidization of particles brings in intensive mixing of plasma gas and solid particles and uniform gas-solid distribution. Therefore, plasma spouted and fluidized beds can serve as excellent reactors for gas-solid and solid-solid reactions, since they combine the advantages of plasma (extremely high temperature and active chemical atmosphere) and the advantages of fluidized and spouted beds (good mixing and the fluidization of the particles). Therefore, problems such as overheating can be dealt with due to the fact that the temperature deviation within the bed is small and operation are very steady, the measured temperature difference is less than 2 K (Goldberger and Oxley 1963).

Also, the presence of solid bed can quench the plasma rapidly and then Fluidized bed reactors are especially advantageous in quenching the solid products and rapid chemical reaction due to the intensive mixing.

On the other hand, the proposal of plasma fluidized bed offers a large range of original properties for the development of new chemical reactors. To illustrate, the electron temperature decreases and the plasma density increases with fluidizing velocity, therefore, active chemical atmosphere can be effectively generated for efficient chemical process. The particles are slightly positively or negatively charged by plasma and therefore the fluidity of the particles can be changed (Mochizuki et al. 1993). On the other hand, moving solid particles can act as catalytic position for the energetic species and therefore increase the utilization of active species and the reaction rate. Also, discharge in the fluidized bed zone is very uniform and stable (El-Naas et al. 1998; Kogelschatz et al. 1997).

Thus, the combination of fluidized/spout bed along with plasma opens up new vistas in gas-solid chemical reaction processes.

Plasma fluidized bed takes advantage of many properties of conventional gas fluidized bed reactors, for example, well-mixed gas-solids interaction, isothermal operation, good temperature control, and fluid-like properties of particles.

Plasma fluidized bed is unique because it involves contact of solids with the active or current-carrying portion of the plasma in a fluidized bed. Therefore, enhanced heat transfer and mass transfer can be achieved inside the one system. Also, discharge in the fluidized bed zone is very uniform and stable. All these features are of great importance for gas-solid process and broaden the occupation of plasma fluidized bed in industrial application with the following advantages (El-Naas et al. 1998; Zhu et al. 1995; Emome et al. 1999; Vaidyanathan et al. 2007; Matsukata et al. 1992; Liu et al. 1996):

- The extremely high temperature, high energy density and high chemical reactivity respectively associated with thermal plasma and non thermal plasma support fast reaction operation, which enables a large throughput from a relatively small-scale reactor;
- Steep temperature and thermal gradients can be controlled independently by plasma chemistry;
- Steep temperature and thermal gradients helps quenching the species rapidly, also the high chemical reactivity leads to rapid reformation, thereby exhausting of persistent organic pollutants (POPs);
- By controlling the resident time, specific gaseous and solid products can be obtained;
- Steady conditions can be reached within very short time, allowing faster start-up and shutdown operation than other thermal treatments such as incineration, without compromising refractory performance;
- Special discharge characteristics inside the reactor allows much more efficient utilization of active species generated in plasma and the input energy;
- Special chemical atmosphere and physical transportation allows different reactions and products to be generated compared with other chemical reactors.

1.4   Potential Excellent Application of Plasma Fluidized Bed

Based on the combined characteristics of plasma process with fluidized bed, plasma fluidized bed reactor for gas-solid reaction brings in several advantages including enhanced heat and mass transfer, generation of high bulk temperature using plasma as a heat source or generation of chemically active species under a much milder condition, strengthening its applications and potential applications. Therefore, plasma fluidized beds are expected to be potential for the many application fields relative to gas-solid chemical process, such as metallurgy extraction, green energy production, environmental protection and advanced materials processes, in which plasma fluidized bed may show surprising performance.

1.5   Evaluating Parameters

For the overall evaluation of the plasma fluidized bed, various parameters should be the taken care of, which can be divided into three groups as:

(1) Parameters of plasma discharge, parameters of electrodes and geometrical configuration, flow parameters (velocity, pressure, carrier gas); voltage-current parameters; temperature-power parameters;
(2) Parameters of fluidized bed: particles parameters (size, chemical characteristics, density, shape), configuration parameters of bed, parameters of gas inlet and outlet, parameters of heat transfer and mass transfer, temperature distribution;
(3) Chemical reactions parameters: exhausted gas composition and gas or solid product composition.

The roles of these parameters on the performance are so important that the investigations of these parameters are of vital role to further study the working process and mechanism and to evaluate and improve the performance of plasma fluidized bed.

References

Allman RM, Walker JM, Hart MK, Laprade CA, Noel LB, Smith CR. Air-fluidized beds or conventional therapy for pressure sores: a randomized trial. Ann Intern Med. 1987;107(5):641–8.
Arpagaus C, Sonnenfeld A, Rohr PRV. A downer reactor for short-time plasma surface modification of polymer powders. Chem Eng Technol. 2005;28(1):87–94.
Attri P, Arora B, Choi EH. Utility of plasma: a new road from physics to chemistry. RSC Adv. 2013;3(31):12540–67.
Christodoulatos C, Korfiatis G, Crowe R, Kunhardt EE. Segmented electrode capillary discharge, non-thermal plasma apparatus and process for promoting chemical reactions. Unite State patent 6818193. 2004.

El-Naas M, Munz RJ, Ajersch F. Modelling of a plasma reactor for the synthesis of calcium carbide. Can Metall Q. 1998;37(1):67–74.

Emome A, Jurewize T. Fuel synthesis for solid oxide fuel cells by plasma spouted bed gasification. 14th International Symposium on Plasma Chemistry, Prague, The Czech Republic. 1999.

Flamant G. Plasma fluidized and spouted bed reactors: an overview. Pure Appl Chem. 1994;66 (6):1231–8.

Goldberger WM, Oxley JH. Quenching the plasma reaction by means of the fluidized bed. AIChE J. 1963;9(6):778–82.

Heberlein J, Murphy AB. Thermal plasma waste treatment. J Phys D Appl Phys. 2008;41 (5):053001.

Kogelschatz U, Eliasson B, Egli W. Dielectric-barrier discharges: principle and applications. J Phys IV. 1997;07(C4):44–66.

Laroussi M. Low-temperature plasmas for medicine? IEEE T Plasma Sci. 2009;37(6):714–25.

Liu LX, Rudolph V, Litster JD. A direct current, plasma fluidized bed reactor: its characteristics and application in diamond synthesis. Powder technology. 1996;88(1):65–70.

Ma JC, Zhao HB, Guo L, Zheng CG. Investigations on batch preparation of iron-based oxygen carrier by spouted bed and using in chemical looping combustion of coal. J Eng Thermophys. 2013;34(10):1960–3.

Mathur KB, Gishler PE. A technique for contacting gases with coarse solid particles. AIChE J. 1955;1(2):157–64.

Matsukata M, Oh-hashi H, Kojima T, Mitsuyoshi Y, Ueyama K. Vertical progress of methane conversion in a DC plasma fluidized bed reactor. Chem eng sci. 1992;47(9-11):2963–2968.

Matsumoto S, Hino M, Kobayashi T. Synthesis of diamond films in a RF induction thermal plasma. Appl Phys Lett. 1987;51(10):737–9.

Mochizuki Y, Ono S, Teii S, Chang JS. Fluidization and plasma characteristics of medium pressure RF glow discharge plasma fluidized bed reactors. Adv Powder Technol. 1993;4 (3):159–67.

Nehra V, Kumar A, Dwivedi H. Atmospheric non-thermal plasma sources. Int J Eng. 2008;2 (1):53–68.

Nezu A, Morishima T, Watanabe T. Thermal plasma treatment of waste ion-exchange resins doped with metals. Thin Solid Films. 2003;435(1–2):335–9.

Pfender E. Thermal plasma technology: where do we stand and where are we going? Plasma Chem Plasma P. 1999;19(1):1–31.

Renzo AD, Maio FPD. Homogeneous and bubbling fluidization regimes in DEM-CFD simulations: hydrodynamic stability of gas and liquid fluidized beds. Chem Eng Sci. 2007;62(1–2):116–30.

Şen Y, Bağcı U, Güleç HA, Mutlu M. Modification of food-contacting surfaces by plasma polymerization technique: reducing the biofouling of microorganisms on stainless steel surface. Food Bioprocess Tech. 2012;5(1):166–75.

Taylor PR, Pirzada SA. Thermal plasma processing of materials: a review. Adv Perform Mater. 1994;1(1):35–50.

Thorley B, Saunby JB, Mathur KB, Osberg GL. An analysis of air and solid flow in a spouted wheat bed. Can J Chem Eng. 1959;37(5):184–92.

Vaidyanathan A, Mulholland J, Ryu J, Smith MS, Circeo LJ. Characterization of fuel gas products from the treatment of solid waste streams with a plasma arc torch. J Environ Manage. 2007;82 (1):77–82.

Yamamoto T. VOC decomposition by nonthermal plasma processing—a new approach. J Electrostat. 1997;42(1–2):227–38.

Zhu CW, Zhao GY, Hlavacek V. A dc plasma-fluidized bed reactor for the production of calcium carbide. J mater sci. 1995;30(9):2412–2419.

Chapter 2
Thermal Plasma Fluidized Bed

Abstract In this chapter, the thermal plasma fluidized bed is introduced in detail. The thermal plasma fluidized bed includes: DC plasma jet spouted bed, AC plasma jet fluidized bed, radio frequency discharge (RF) fluidized bed (RF plasma fluidized bed, RF plasma circulating fluidized bed, RF downer bed), microwave discharge fluidized bed and electrothermal plasma fluidized bed. Moreover, this chapter introduces the research progress and applications of various reactors, and points out the shortcomings of these reactors.

Keyword Thermal plasma fluidized bed

2.1 DC Plasma Jet Spouted Bed

Generally, plasma spouted bed reactors (PSBR) refer reactors in which the plasma jet is used to offer fully gas-solid contact, to heat up the solid particles and to provide active species, including the excited species, ionized species and active molecules (O_3 and NO_x) to react with the particles.

Generally operated under low voltage and high current conditions, DC plasma torches are able to offer extremely high temperature atmosphere and higher energy density, showing great potential for generating very high temperatures without combustion, and may offer several advantages over conventional combustion technology. Generally, plasma jet can be generated by the induction of carries gas through the torch connected with high voltage, and this principle is similar to that of spouted bed. So, DC plasma torch reactor and the spouted catalytic bed have similar reactor configurations and flow mechanisms, depending on the feed flow. Therefore, the DC plasma torch spouted bed can be expected to be an ideal choice for the highly endothermic gas–solid and solid–solid reactions (El-Naas et al. 1998b; Zhu et al. 1995).

Since 1963 when Goldberger and Oxley proposed the first DC plasma fluidized bed for rapid quenching of particles in extremely high temperature, DC torch plasma fluidized bed has been extensively investigated by researchers due to its

© Springer Nature Singapore Pte Ltd. and Zhejiang University Press 2018

C. Du et al., *Plasma Fluidized Bed*, Advanced Topics in Science and Technology in China, https://doi.org/10.1007/978-981-10-5819-6_2

unique advantages in super high temperature operation. Therefore it exited various types of DC torch fluidized bed for different applications (El-Naas et al. 1998a, b; Zhu et al. 1995; El-Naas 1996; Emome and Jurewize 1999; Vaidyanathan et al. 2007; Matsukata et al. 1994; Liu et al. 1996a, b; Bal et al. 1971; Flamant and Bamrim 2000a, b; Tsukada et al. 1995).

As seen in Fig. 2.1a and b, DC torch plasma fluidized/spouted bed can be divided into two categories, based on the ways by which the torch is located on the fluidized bed reactor. For reactors shown in Fig. 2.1a, the plasma torch is proposed in order to avoid the possible dropping of the powder into the bed through the holes on the top of distributor upon the plasma torch, which can in turn disturb the electrical discharge and even cause the damage of the electrodes. The plasma gas is passed through the nozzle which is located on the side wall of the fluidized bed (Flamant and Bamrim 2000a, b).

Actually, the much more utilized reactor was the type in Fig. 2.1b. The plasma jet is used as the vector for particle circulation, as heat source and as reacting fluid medium. Many researchers especially Mass et al. have made an endeavor to developing and utilizing such a kind of DC torch fluidized bed (El-Naas et al. 1995, 1998a, b). Taking Mass's reactor for example, the plasma torch was composed of a conical thoriated tungsten cathode and an annular copper anode, with a design power of the order of 20 kW (El-Naas et al. 1998a). Plasma gases were injected into

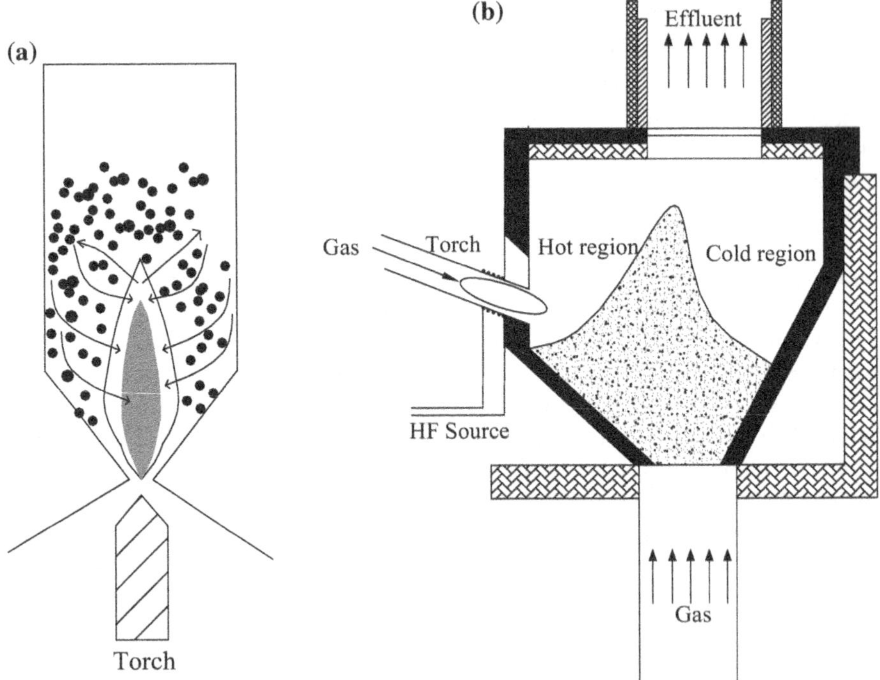

Fig. 2.1 Two categories of DC torch plasma fluidized/spouted bed

the system tangentially and passed through the gap between the cathode and anode. To reduce the turbulence caused by the plasma jet and ensure the stability of the plasma arc, a large swirling velocity is located through four screens (El-Naas et al. 1998a). For a DC plasma torch fluidized bed, the design of the distributor is of vital importance for the stable and continual operation of the bed reactor to ensure efficient fluidization and avoid the possible overheating of reactors. Also, the possible enlargement of the small holes on the distributor is another problem needs considering. A typical well designed distributor was designed by Zhu et al. (Zhu et al. 1995). Also the design of the base geometry is very important for the performance of DC torch plasma fluidized bed. With both batches, the interaction of the plasma jet with the granular material is limited to a central zone in the reactor, the extent of which is more important for coarse materials than fine materials (Tsukada et al. 1995).

A plasma torch fluidized bed can be divided into two different zones: the reaction zone at the bottom of bed zone where most particles were in the fluidized state (El-Naas et al. 1998a). In the reaction zone, the temperature is much higher than that of the bed zone and the size of the plasma zone depends on the plasma condition and is proportional to the plasma jet enthalpy. In the fluidized bed, particles circulate from the bed into the reaction zone and react when immersed in high temperature up to 6000 K, which is dependent on the power input and the addition of hydrogen. Circulation of the particles in the whole bed reactor finally results in the enhancement of the heat transfer between plasma jet and the injected particles and a smaller temperature difference in the reaction zone to ensure the constant surface temperature of the particles.

Thus, such a fluidized bed has been investigated for the various metallurgy extraction processes such as reduction of sulphide, separation of mineral oxides, and reduction of oxides and conversion of volatile chlorides to metals. Also, the synthesis of the calcium carbide with the DC plasma torch fluidized bed has also been extensively investigated. Experimental result showed that within a plasma torch fluidized bed, the conversion of target product is very high and that the temperature and reaction time required for the production of calcium carbide are much lower than those used in the conventional operation process today. Previous studies indicated that with the combination of highly reactive species with high temperature condition, the DC plasma torch fluidized bed is able to lower the energy consumption up to 40% (El-Naas et al. 1998a). However, the role of the generated active species played on the synthesis process has not yet been determined and still unknown. On the other hand, although the arrangement was effective in the transferring plasma enthalpy to bed materials, it can result in the high temperature sufficiently high, partially reacted particles soften and agglomerate when reaching a temperature near the melting point of carbide. The large particles agglomerates tended to fall into the bottom of jet, where they combined and formed a cylindrical mass around the jet. Such a limitation made it difficult to achieve a higher conversion rate.

To deal with this problem, continuous plasma fluidized bed process and a lower energy density are required so that the high temperature process can be in a much more stable state and that the problem of melting can be expected to be eliminated (Kogelschatz et al. 1997; Gomez et al. 2008).

2.2 AC Plasma Jet Fluidized Bed

Different from universally operated under low-voltage and high-current conditions for dc plasma torches, recently Lee et al. reported a nonthermal plasma torch spouted bed operating at high-voltage and low-current conditions (Lee and Sekiguchi 2011). Figure 2.2 shows the schematic diagram of nonequilibrium plasma torch fluidized/spouted bed. Its discharge power is also far lower than the power often needed in thermal arc plasma systems. After an arc discharge occurring in the vicinity of the smallest inter-electrode gap, the gas flow blows the arc upstream and out of the electrode to form a flame-shaped plasma region, which is particularly suitable for low temperature chemical operation. Comparing with

Fig. 2.2 A detailed representation of the fluid bed reactor

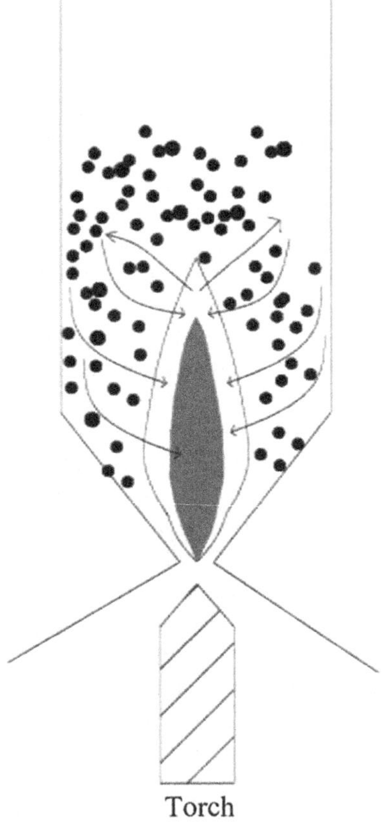

conventional flame technology applied in manufacture, this plasma technology has advantages in decreasing CO_2 and pollutants release, easily selecting raw materials, and ability to synthesize various nanoparticles. By the help of template-controlled growth technique, the reactive nonequilibrium plasma may even be used to synthesize more complex nanostructures such as metal oxide nanowires.

2.3 Radio Frequency Discharge (RF) Fluidized Bed

2.3.1 RF Plasma Fluidized Bed

Low-frequency plasma is sometimes used because generators are less expensive and do not require precise impedance matching. However, at low frequencies plasma may extinguish each half cycle and reaction rates are significantly slower than at RF. Therefore, most plasma reactors use higher frequencies. At 13.56 MHz the plasma is very stable and reactive because the quench time of the plasma species is much longer than the time between half-cycles of the excitation. RF plasma and microwave plasma fluidized beds have become very popular since the 90's and widely investigated for various applications and especially for surface treatment of particulate materials (Lee and Sekiguchi 2011; Schmidt-Szalowski et al. 2006; Ua-amnueychai et al. 2015). RF plasma is easily generated with equipment that is reliable and has been used commercially for many years (Kroker et al. 2012).

There are basically two different kinds of RF plasma fluidized bed based on the ways of RF power coupling: capacitively coupled plasma (CCP) fluidized bed or inductively coupled plasma (ICP) fluidized bed. Inductive coupling is realized by a planar or a helical coil, mostly by external coupling through a dielectric wall around the outside wall of the fluidized bed rather than directly in the plasma zone (Kroker et al. 2012; Chen et al. 2008). A time-variant magnetic field is induced by the RF powered coil, leading to electron acceleration. Being placed at the outer part of the reactor the copper coil prevents the powder to be deposited on internal electrodes. The number of turns of the copper coil of ICP is favorable if a high ion density and low ion bombardment of the adjacent walls is desired. However, compared with CCP fluidized bed, ICP fluidized bed is much less popular used for the particulate processing. RF plasma torches utilize inductive or capacitive coupling to transfer electromagnetic energy from the RF power source to the plasma working gas. The RF plasma originally developed by Reed in 1960 (Chen et al. 2006) as a new crystal-growing technique, received interest in the field of plasma chemistry. They are very compact and deliver extraordinarily high input energy per unit volume. The ability of the RF plasma reactor, owing to the absence of metal electrodes, to handle virtually any chemical, opened up the way to investigations of many chemical processes that could previously not be studied in this way because of the rapid corrosion of plasma torch materials. RF current and microwaves can be transferred through insulators, so the use of external electrodes is possible, and the

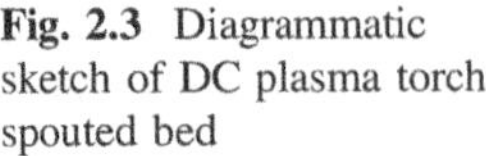

Fig. 2.3 Diagrammatic sketch of DC plasma torch spouted bed

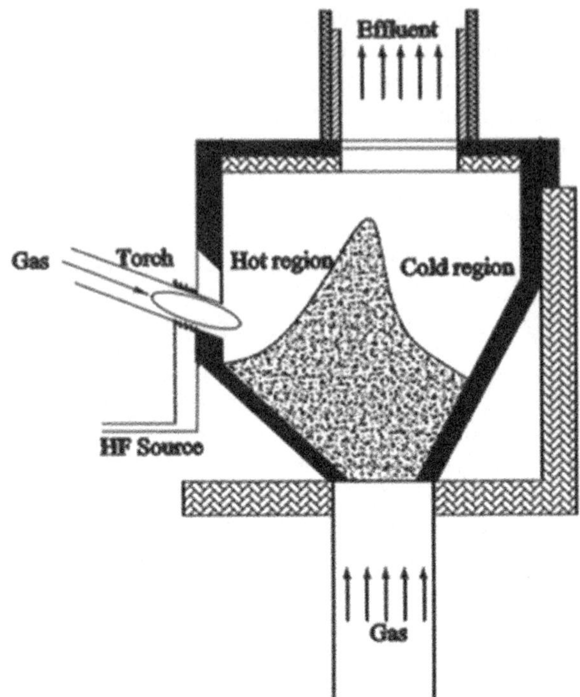

external electrodes can be inductive coil as shown in Fig. 2.3 or capacitive coupling electrode (Kumar et al. 2008; Attri et al. 2013). In this way, the electrodes are not exposed to the severe conditions of thermal plasma and, therefore, have a very long lifetime. RF plasma generators are commonly available at power levels of 100 kW. Scale-up has been demonstrated to the 1 MW range (Du 2014a).

For most plasma processes, including deposition and etching, capacitive coupling is of advantage. Here, electrodes are arranged in parallel plate configuration in the simplest case, where two electrodes are split by a distance of a few centimeters. They may be in contact with the discharge or be insulated from a dielectric. It is also possible to place the electrodes outside the reactor vessel, in the case of insulation chamber walls (e.g. quartz glass). The electrodes are connected to the RF generator via an impedance matching network. The matching network is necessary to match the impedance of the generator to that of discharge.

To stabilize the discharge inside the fluidized bed, a low pressure condition is necessary in need due to the unique properties of low pressure discharges. These include moderate ignition voltages, the ability to create large-volume glow discharges, which have a considerable advantage for homogeneous surface modification, and typically low overall gas temperatures that result from inefficient energy transfer between energetic particles (electrons) and heavy particles (ions, neutrals), extended to temperature sensitive materials (e.g. polymers).

Typical RF plasma fluidized bed configuration can be described as Fig. 2.4 (Bretagnol et al. 2004). The fluidized bed at low pressure is generated by passing a flow of a determined gas through a bed of powder which is placed on porous support. The fluidized bed reactor consists of a cylindrical Pyrex glass tube of 30 mm internal diameter and 600 mm height as shown in Fig. 2.4. Powders were supported on a porous glass plate. A capacitive coupled system composed of external electrodes (distance between the two electrodes: 6 cm), placed at the same level as the fluidized bed were used. Electrodes were coupled with a 13.56 MHz radio-frequency generator connected to a match box. Incident and reflected powers were controlled by a wattmeter (impedance 50 X). A very good contact between reactive species of the plasma discharge and powder surfaces is obtained. However, particles in the C group have not still present many problems and fluidization remains more like an art.

One of the main drawbacks presented by this type of reactor is therefore that in order to have a good fluidization behavior, particles should not present smaller diameters than 40 microns. It should be also pointed out that in order to obtain good fluidization particle size should present a narrow particle distribution in order to avoid stratification of the treated powders.

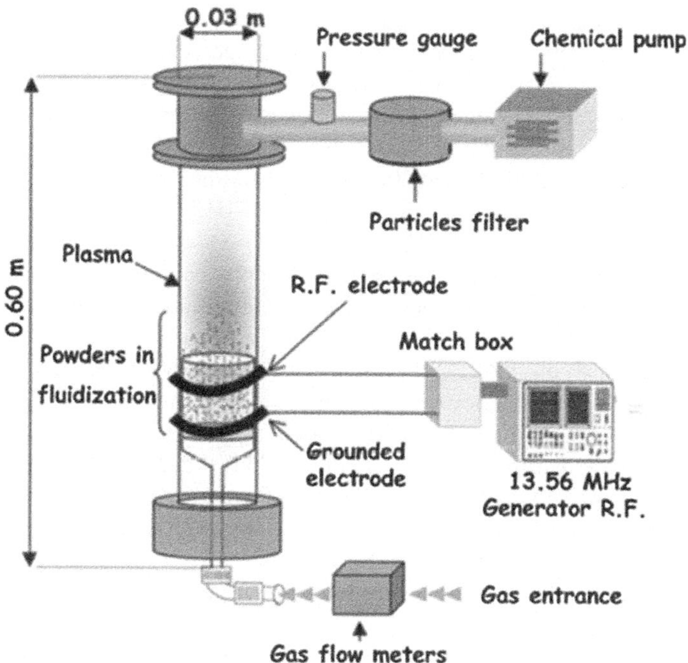

Fig. 2.4 Typical RF plasma fluidized bed configuration

2.3.2 RF Plasma Circulating Fluidized Bed

However, yet, a major limitation of this kind of plasma is the requirement for relatively expensive and complicated vacuum system. From the economic and ecological points of view, there is a big interest to transfer low-temperature plasma surface modification techniques at low pressure to processes at atmospheric pressure. Concerning powder modification using APP techniques, the number of published works is much lower than for LPP systems. The difficulty to handle powders in open air systems and make them flow throw the APP region seems to be the main drawback. However some interesting works should be presented. In continuous processes such as the one presented in this work, the need of powder fluidization to obtain a homogeneous treatment at the plasma zone, as well as the maximal recovery amount of the CB after the modification are two important points that should be studied in detail. Blockage of the system caused by the agglomeration of the filler on the walls should obviously be avoided. Although some systems have been described using plasma at atmospheric pressure they have only been able to treat small amounts of filler using batch reactions and they are briefly here described. The group has been presented as one of the pioneers in the advances in the APG reactor was also one of the first to treat powders using this technique (Prat et al. 2000). They used a circulating APG reactor to treat Silica particles (average diameter 152 µm). In this case the glass tube works as the dielectric layer. Helium was used as plasma gas (3000 sccm) with 1 percent of tetrafluoroethylene (30 sccm). The reactor worked at a frequency of 15 kHz and 10 W of Power. Each batch reaction was able to treat mentioned powder. The same group used the APG reactor to treat Fe_3O_4 pigments with He: TEOS (500 sccm: 10 mg/min) at 13.56 MHz and 250–300 W. In 2001 they used the reactor in order to coat amorphous magnetic powders (Co70.3 Fe4.7 Si10 B15) with zirconia ZrO_2 operating at RF 13.56 MHz and 400 W. A mixture He: O_2 (6500 sccm: 15 sccm) was utilized for this purpose. In order to obtain the ZrO_2 coating, butyl zirconate was previously adsorbed on the magnetic powder which had an average diameter of 50 µm (Liu et al. 1996a, b).

Other groups also used the APP plasma circulating reactors to modify powders. This is the case of Alumina powders (60 µm) coating in a RF reactor (13.56 MHz) by Park SM et al. (Park et al. 2005). In this case a powder eight of about 50 grams was used for each experiment. A mixture of He, Ar, O_2 and TEOS with a total flow of 1200 sccm was used. In the all the described experiments till to this point, the glow state of the discharge was achieved by using high flows of Helium which consequently made the treatment quite expensive. As a consequence, this new method was seen as a great opportunity for large scale surface modification at low processing costs.

Fast fluidized flow is generally characterized by a diluted upward solid flow in the tube center and a fluctuating downward flow of strands of particles at the tube wall. This intense longitudinal back-mixing gives rise to a flat vertical temperature profile. The coated particles are entrained in the flowing gas and thus removed from the plasma zone, thereby dissipating the plasma-generated heat from the reactor, subsequently separated from the gas in a cyclone and externally recirculated through the down comer and L-valve. The particle concentration in the riser is proportional to the circulation rate, which is adjusted by the aeration gas flow rate into the L-valve. The gas exit of the cyclone is connected to the vacuum unit.

One of the main disadvantages of the circulating reactor is that particles should be in the hundreds of micron scale, as separation of powder and gas would be too difficult in the nanometric region. There is no evidence that fluidized neither circulating systems have been used in order to modify carbon black, probably the size and the electrostatic properties are the main drawbacks to use such type of reactors to modify this powder (Liu et al. 1996a, b; Bal et al. 1971; Flamant and Bamrim 2000a, b; Tsukada et al. 1995).

The plasma stability of particle–plasma interaction can be characterized by visual examination. The plasma brightness and the glow volume are used as criteria (see Fig. 2.4). The plasma stability is insensitive to gas composition. A small amount of O_2 reduces the brightness, but the plasma glow is still stable. The introduction of particles alters the glow appearance in color and intensity. The preliminary experimental tests reveal that the process stability can be increased by a higher RF power, lower pressure, lower gas flow rate and a lower amount of O_2 in the process gas. Despite the advantageous features of the downer plasma treatment, the existing reactor setup suffers from the adhesion of polymer particles at localized zones of the reactor wall. This drawback limits the maximum processing time of the process to a few minutes (Du 2014b). It is considered that the deposits are formed due to electrostatic loadings. If the polymer particles are injected into the plasma, they become negatively charged. This undesirable side effect so far hinders the plasma-assisted treatment from becoming the method of substitution for wet-chemical surface activation of polymer powders. Elimination of these deposition phenomena is a substantial objective of ongoing research work because a continuous processing would be necessary for industrial-scale application.

2.3.3   RF Downer Reactor

It is thought to be worthy to mention is the one designed by the group of Aspagaus et al. in 2005 (Arpagaus et al. 2005a, b). This reactor could remind the above presented but working in the opposite direction. In this case the powder is placed in a reservoir at the top of the reactor. The powder is fed through a metering screw which pushes the powder to go down to the reactor bottom while passing through the plasma zone. The reactor operates at RF and two half shell copper electrodes, fit to the outer shape of the reactor, capacitively couple the plasma. HDPE with an average diameter of 56 µm and a density of 950 kg/m^3 were modified by means of O_2/Ar plasma (Cormier and Rusu 2001; Ye et al. 2004). The powder was in contact with the plasma less than 0.1 s and the powder throughput was kept at 5 kg/hr while operation times were between 1 to 5 min depending on the amount of modified powder desired.

2.4 Microwave Discharge (MW) Fluidized Bed

The principle of excitation of plasma by microwaves is similar to the excitation with RF. Some differences, however, result from the higher frequency. Moreover, the application system is different. Due to the higher frequency compared to RF, the maximum velocity an electron can reach in a MW field is smaller and therewith the maximum energy acquired during one cycle is smaller too. In a collisionless situation at 2.45 GHz it reaches about 0.03 eV (El-Naas et al. 1998a, b). Often the plasma volume is separated from the applicator by a quartz glass or other dielectric window. MW-generated plasma generally has higher electron energies than RF-generated plasmas, typically ranging from 5 to 15 eV rather than from 1 or 2 eV characteristic of the latter discharges. MW plasma does not have a high voltage sheath, with its accompanying ion sputtering of the walls. On the other hand, substrate biasing is not possible with MW.

Different heating methods for fluidized beds have been investigated up to now. These methods can be divided into indirect and direct heating of the bed.

The common method of indirect heating is resistance heating of the reactor walls. If the heat input from the surface of the reactor is not sufficient, a heat exchanger can be installed directly in the fluidized bed, although hydrodynamic and abrasion behavior is influenced negatively. Particularly for drying processes pre-heating of the fluidization gas is applied. However, if the fluidization gas is hotter than the particles itself, reactive gasses needed for chemical reaction would undergo homogeneous gas phase reaction instead of heterogeneous transfer and chemical reaction at the particle surface.

Most important direct heating methods for processes with large mass flow exothermic reactions are used, which release energy directly into the fluidized bed in order to reach a temperature level not achievable economically by indirect heating methods. The best example is coal gasification, where partial oxidation is executed, which however limits the selection of the gas atmosphere and the yield. In case of chemical reaction only highly exothermic reactions could be used for direct heating, because of dilution of the chemical reaction gases with the fluidization gas. Other methods of direct bed heating use dielectric loss, e.g., capacitive heating, microwave or RF-heating.

The main difference between direct and indirect heating is the direction of heat flux: when heating the reactor wall or the gas, the heat has to be transferred from the wall to the fluid and/or from the fluid to the particles. Direct heating means heat flux from the particles to the fluid and to the walls. In both cases temperature gradients are the driving force for the overall heat flux. Although on principle direct dielectric heating of the bed should be possible without a macroscopic temperature gradient, practically due to the heat losses through the wall an inverse temperature gradient may develop between the bed and the walls of the FBR. Microwave heating of fluidized beds requires sophisticated solutions for feeding high power microwaves into the FBR, due to the dusty environment inside the reactor. A useful approach to solve this problem in a pilot plant installation with 75 kW magnetrons operating at 915 MHz frequency using wave guides has been presented by the Canadian company EMR, for roasting ores in a microwave heated fluidized bed reactor since 1999 (El-Naas et al. 1998a, b; Pacek and Nienow 1990; Kono et al. 1987; Visser 1989; Rogers and Morin 1991; Prat et al. 2000).

However, successful scale up of microwave heated FBR processes requires a thorough investigation of microwave specific parameters, like e.g., penetration depth of the radiation into the fluidized bed, plasma ignition at high microwave power levels due to pressure variations and electrostatic charging, and reliability of microwave coupling into the dusty environment of a fluidized bed reactor (Chang et al. 1987; Feng et al. 2002). No general solution can be provided, and rather an optimized solution should be developed for each particular process. Sensitivity to temperature dependent dielectric and discharge properties has to be fully implemented into the process control, in order to arrive at a successful microwave heated FBR's on an industrial scale. This novel intensive process offers the potential for quick and simple production of functional composite particles.

Nevertheless, as underlined by Karches et al. (1999), Karches and Rohr (2001), and Karches et al. (2004), a major drawback of putting together in contact cold plasma and particles is the resulting heating (formation of hot points), which is unacceptable for thermally sensitive materials. An interesting way to overcome this impediment has been developed by Mutel et al. (2004), the fluidized bed is placed 0.65 m downstream a microwave nitrogen discharge. This allows avoiding the contact between the plasma energetic species (electrons, ions, etc.) and the powders which is responsible for the powder bed overheating. However, this cannot be achieved with any gas. Nitrogen-plasma generates long-living chemically active species, mainly atomic nitrogen and metastable species, and so, active post-discharges (Heberlein and Murphy 2008; Pajkic and Willert-Porada 2009). This novel intensive process offers the potential for quick and simple production of functional composite particles. It enables simplification and cost reduction of the fluidized bed coating process, because no dangerous, expensive and difficult-to-handle metal precursors have to be used, and the off-gas treatment is also much easier. Further cost reductions include also elimination of the complicated and expensive pumping and trapping subsystems, since no vacuum is needed for this process.

2.5 Electrothermal Plasma Fluidized Bed

When electricity is passed through a fixed bed of electrically conducting particulate solids, the bed offers resistance to the flow of current; this resistance depends on many parameters, including the nature of the solid, the nature of the linkages among

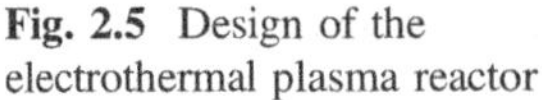
Fig. 2.5 Design of the electrothermal plasma reactor

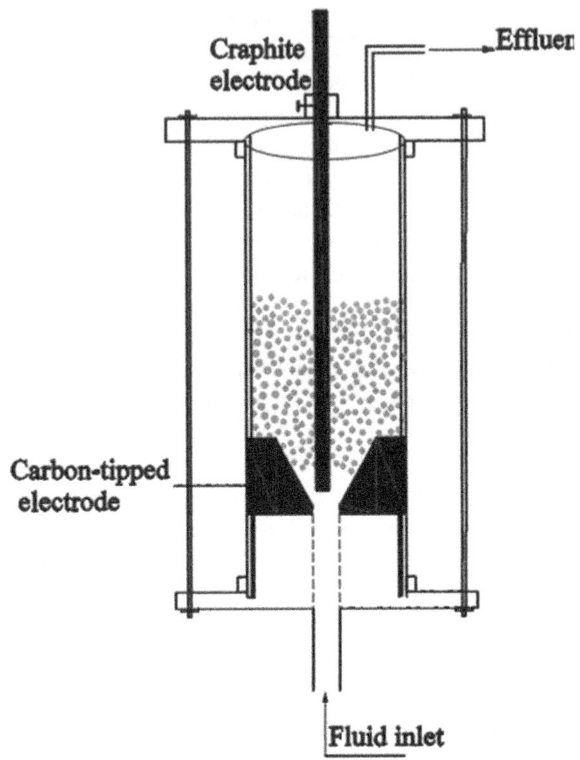

the particles within the bed, the bed voidage, the bed height, the electrode geometry, etc. If the same fixed bed is fluidized by passing a gas, the resistance of the bed increases; the resistance offered by the conducting particles generates heat within the bed and can maintain the bed in an isothermal condition (Takarada et al. 1993). A fluidized bed thus heated by the passage of current through the bed of conductive particulate solids is termed an electrothermal fluidized bed or electrofluid reactor. A typical electrothermal fluidized bed reactor is depicted in Fig. 2.5 (Steinbach 1996).

Granular carbon is a good electrical conductor providing a surface that catalyzes chemical reduction and provides deposition sites for carbon and other elements. The bench scale electrothermal uses an electric arc current coupled with a fluidized carbon-particle bed to generate the plasma environment (Fig. 2.5) which has been designed by Bashlai et al. (Bashlai et al. 1972) and investigated by Steinbach (Steinbach et al. 2003). This design allows for a low electrical current (2–5 A) and a high in-flow of gases. Carbon particles can provide deposition sites for carbon and other elements. The carbon is fluidized with gas to prevent electrical shorting between the electrodes. This fluidization is accomplished by injecting influent gases into the bottom electrode and out through three small holes bored into the cup-shaped carbon tip.

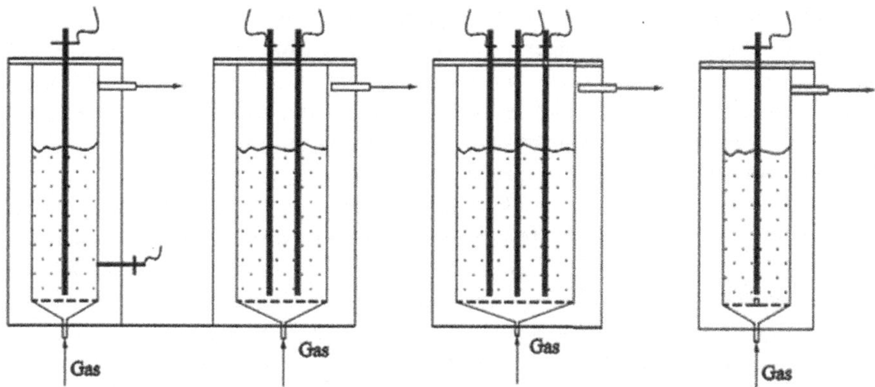

Fig. 2.6 Configurations of electrothermal fluidized bed and electrode arrangements

Schematics of various configurations of the electrothermal fluidized bed reactor are shown in Fig. 2.6 (Steinbach 1996). In Fig. 2.6, the reactor vessel or the fluidizing column itself serves as one of the two electrodes; the other electrode is centrally suspended from the top of the reactor and submerged inside a fluidized bed of conducting solids. As the electrothermal fluidized bed operates at high temperatures (up to 4500 °C), it is essential to have good insulation and also to select a proper high temperature resistant reactor vessel. For most experimental or large-scale reactors, graphite is chosen as construction material and the column is generally insulated with lampblack and suitable refractory granules. Additional insulation between the outer metallic jacket and the refractory granules is provided with materials such as sovlit crumb. In a typical experimental 100 mm diameter graphite column electrothermal fluidized bed with a concentrically suspended graphite electrode (Steinbach 1996), the outer insulation provided consists of 8 mm of lampblack and 100 mm of refractory granules. In order to reduce radiation heat losses, radiation shields in the form of graphite plates above the bed and graphite central electrodes may be used.

Electrofluid reactors have potential application in many chemical, metallurgical, and metalworking industries. In many high-temperature reactions, electrofluid reactors offer in situ heating during the reaction and the heating is stopped automatically when the charge is consumed. In other words, these electrothermal fluidized bed reactors for chemical reactions save energy because no external heating or transfer of heat is required. An externally mounted furnace, especially for high-temperature reactions with gases/halogens, requires careful maintenance and regular replacement of the heating elements. In an electrofluid reactor, such problems are totally eliminated. A mandatory requirement for a reactor of this type is a bed of solid that is electrically conducting and does not volatilize during heating. Thus, it appears that the reactor is of no use if nonconducting solids are to be heated. This problem can be overcome by mixing a nonconducting solid with a conducting solid. This principle of heating and reacting minerals in an electrothermal fluidized bed is used in pyrometallurgical operations, and the relevant details will be presented later as part of the discussion on direct chlorination of refractive materials/silicate ores such as zircon. An earlier major industrial application of the electrothermal fluidized bed reactor was in the production of hydrogen cyanide (Wu et al. 2010; Fridmari et al. 1997; Hu et al. 2015; Gerdes et al. 2006; Fauchais and Vardelle 2000; Lee et al. 1985; Rykalin 1976; Waldie 1972; Chen et al. 1995; Chen and Pfender 1983a, b; Uglov and Gnedovets 1991). By using an electrothermal fluidized bed, a faster reaction at high temperature was achieved without using the costly platinum catalyst which otherwise is essential for this reaction. Application of this reactor was subsequently reported in many other fields. The electrothermal fluidized bed has been used in coal industries mainly for the gasification of coal char (Yang et al. 2006) in the production of synthesis gas which can be transported through pipelines both conveniently and economical.

All in all, for thermal plasma fluidized beds, relatively high temperature can offer energy needed for chemical reaction together with the reactive species of high density for the activation of the materials and lower the energy consumption, to illustrate, for application of metallurgy process, active species and enhanced transfer inside the dc plasma torch fluidized bed can effectively make the extraction temperature much lower than that of conventional thermal operation. As for RF and microwave fluidized bed, with the electrodes outside the fluidized zone, the electrodes can be kept away from erosion, much more and stable species can be generated, which is of great importance for chemical application.

References

Arpagaus C, Rossi A, Von Rohr PR. Short-time plasma surface modification of HDPE powder in a plasma downer reactor-process, wettability improvement and ageing effects. Appl Surf Sci. 2005a;252(5):1581–95.

Arpagaus C, Sonnenfeld A, Von Rohr PR. A downer reactor for short-time plasma surface modification of polymer powders. Chem Eng Technol. 2005b;28(1):87–94.

Attri P, Arora B, Choi EH. Utility of plasma: a new road from physics to chemistry. RSC Adv. 2013;3(31):12540–67.

Bal S, Musialski A, Swierczek R. Gasification of coal fines in a laboratory plasma-chemical reactor with a spouted bed. [Ar plasma]. Vet Rec. 1971;147(6):166–7.

Bashlai KI, Barantsev IF, Grinbaum MB, Stanyakin VM, Samodurov VV, Todes OM. Thermal and electrical characteristics of a high-frequency electrothermal fluidization bed. J Eng Phys. 1972;22(6):665–9.

Bretagnol F, Tatoulian M, Arefi-Khonsari F, Lorang G, Amouroux J. Surface modification of polyethylene powder by nitrogen and ammonia low pressure plasma in a fluidized bed reactor. React Funct Polymers. 2004;61(2):221–32.

Chang JS, Ono S, Teil S. Medium pressure glow discharge plasma oxidation by fluifized bed reactors, 1987.

Chen X, Pfender E. Effect of the Knudsen number on heat transfer to a particle immersed into a thermal plasma. Plasma Chem Plasma P. 1983a;3(1):97–113.

Chen X, Pfender E. Behavior of small particles in a thermal plasma flow. Plasma Chem Plasma P. 1983b;3(3):351–66.

Chen X, Chen J, Wang Y. Unsteady heating of metallic particles in a rarefied plasma. Plasma Chem Plasma P. 1995;15(2):199–219.

Chen G, Chen S, Zhou M, Feng W, Gu W, Yang S. Application of a novel atmospheric pressure plasma fluidized bed in the powder surface modification. J Phys D Appl Phys. 2006;39 (24):5211.

Chen G, Chen S, Feng W, Chen W, Yang SZ. Surface modification of the nanoparticles by an atmospheric room-temperature plasma fluidized bed. Appl Surf Sci. 2008;254(13):3915–20.

Chen G, Zhou M, Chen S, Lv G, Yao J. Nanolayer biofilm coated on magnetic nanoparticles by using a dielectric barrier discharge glow plasma fluidized bed for immobilizing an antimicrobial peptide. Nanotechnology. 2009;20(46):465706.

Cormier JM, Rusu I. Syngas production via methane steam reforming with oxygen: plasma reactors versus chemical reactors. J Phys D Appl Phys. 2001;34(34):2798.

Du C. A plasma fluidized bed for the production of syngas from MSW. China patent 201410844203.9. 2014a.

Du C. A plasma fluidized bed for the cineration of fly ash. China patent 201410850031.6. 2014b.

El-Naas MH, Munz R, Ajersch F. Modelling of a plasma reactor for the synthesis of calcium carbide. CANMetallQuart. 1998a;37(1):67–74.

El-Naas MH, Munz R, Ajersch F. Solid-phase synthesis of calcium carbide in a plasma reactor. Plasma Chem Plasma P. 1998b;18(3):409–27.

EL-Naas MH, Munz R, Ajersch F. Production of calcium carbide in a plasmajet fluid bed reactor. Proc ISPC-12. 1995:613–8.

El-Naas MH. Synthesis of calcium carbide in a plasma spout fluid bed. Montreal: McGill University; 1996.

Emome A, Jurewize T. Fuel synthesis for solid oxide fuel cells by plasma spouted bed gasification. In: 14th international symposium on plasma chemistry (Prague, 1999); 1999.

Fauchais P, Vardelle A. Pending problems in thermal plasmas and actual development. Plasma Phys Controlled Fusion. 2000;42(12B):B365.

Feng H. Analysis of microwave assisted fluidized-bed drying of particulate product with a simplified heat and mass transfer model. International Communications in Heat & Mass Transfer, 2002;29(8):1021–1028.

Flamant G, Bamrim A. The plasma spouted bed reactor for applications in metallurgy and material synthesis. High Temp Mater Process. 2000a;4(4):455–71.

Flamant G, Bamrim A. The plasma spouted bed reactor for applications in metallurgy and material synthesis. High Temp Mater Process. 2000b;4(4):18.

Fridmari HSA, Saveliev A, Nester S, Kerirzedy L. Nonequilibrium gliding arc in fluidized bed. In: 13th international symposium on plasma chemistry (Beijing, 1997). 1997.

Gerdes T, Tap R, Bahke P, Willert-Porada M. CVD–processes in microwave heated fluidized bed reactors. Adv Microwave Radio Freq Process. 2006;54–55(09):720–34.

Gomez E, Rani DA, Cheeseman CR, Deegan D, Wise M, Boccaccini AR. Thermal plasma technology for the treatment of wastes: a critical review. J Hazard Mater. 2008;161(4):614–26.

Heberlein J, Murphy AB. Thermal plasma waste treatment. J Phys D Appl Phys. 2008;41 (5):053001.

Hu MB, Dang SC, Ma Q, Xia WD. Stabilizing effect of plasma discharge on bubbling fluidized granular bed. Chin Phys B. 2015;24(7):288–92.

Karches M, Rohr PRV. Microwave plasma characteristics of a circulating fluidized bed-plasma reactor for coating of powders. Surf Coat Tech. 2001;142–144(3):28–33.

Karches M, Bayer C, Rohr PRV. A circulating fluidised bed for plasma-enhanced Chem vapor deposition on powders at low temperatures. Surf Coat Tech. 1999;116–119(4):879–85.

Karches M, Takashima AH, Kanno Y. Development of a circulating fluidized-bed reactor for microwave-activated catalysis. Ind Eng Chem Res. 2004;43(26):8200–6.

Kogelschatz U, Eliasson B, Egli W. Dielectric-barrier discharges. Principle and applications. J Phys IV. 1997;7(C4):44–66.

Kono HO, Soltani-Ahmadi A, Suzuki M. Kinetic forces of solid particles in coarse particles fluidized beds. Powder Technol. 1987;52(1):49–58.

Kroker T, Kolb T, Schenk A, Krawczyk K, Młotek M, Gericke KH. Catalytic conversion of simulated biogas mixtures to synthesis gas in a fluidized bed reactor supported by a DBD. Plasma Chem Plasma P. 2012;32(3):565–82.

Kumar A, Dwivedi HK, Nehra V. Atmospheric non-thermal plasma sources. International J. Eng, 2008;2(1):53–68.

Lee H, Sekiguchi H. Plasma-catalytic hybrid system using spouted bed with a gliding arc discharge: CH4 reforming as a model reaction. J Phys D Appl Phys. 2011;44(27):274008.

Lee YC, Chyou YP, Pfender E. Particle dynamics and particle heat and mass transfer in thermal plasmas. Part II. Particle heat and mass transfer in thermal plasmas. Plasma Chem Plasma P. 1985;5(4):391–414.

Liu L, Rudolph V, Litster J. A direct current, plasma fluidized bed reactor: its characteristics and application in diamond synthesis. Powder Technol. 1996a;88(1):65–70.

Liu LX, Rudolph V, Litster JD. A direct current, plasma fluidized bed reactor: its characteristics and application in diamond synthesis. Powder Technol. 1996b;88(1):65–70.

Matsukata M, Suzuki K, Ueyama K, Kojima T. Development of a microwave plasma-fluidized bed reactor for novel particle processing. Int J Multiph Flow. 1994;20(4):763–73.

Morstein M, Karches M, Bayer C, Casanova D, Rohr PRV. Plasma CVD of ultrathin TiO_2 films on powders in a circulating fluidized bed. Chem Vapor Depos. 2000;6(1):16–20.

Mutel B, Bigan M, Vezin H. Remote nitrogen plasma treatment of a polyethylene powder: optimisation of the process by composite experimental designs. Appl Surf Sci. 2004;239 (1):25–35.

Pacek AW, Nienow A. Fluidisation of fine and very dense hardmetal powders. Powder Technol. 1990;60(2):145–58.

Pajkic Z, Willert-Porada M. Atmospheric pressure microwave plasma fluidized bed CVD of AIN coatings. Surf Coat Tech. 2009;203(20):3168–72.

Pajkic Z, Wolf H, Gerdes T, Willert-Porada M. Microwave plasma fluidized bed arc-PVD coating of particulate materials. Surf Coat Tech. 2008;202(16):3927–32.

Park SH, Sang DK. Functionalization of HDPE powder by CF_4 plasma surface treatment in a fluidized bed reactor. Korean J Chem Eng. 1999;16(6):731–6.

Park SM, Jung SH, Park SH, Kim SD. Silicon oxide thin film deposition on alumina in a circulating fluidized bed reactor. Key Eng Mater. 2005;277:577–82.

Prat R, Koh YJ, Babukutty Y, Kogoma M, Okazaki S, Kodama M. Polymer deposition using atmospheric pressure plasma glow (APG) discharge. Polymer. 2000;41(20):7355–60.

Rogers T, Morin TJ. Slip flow in fixed and fluidized bed plasma reactors. Plasma Chem Plasma P. 1991;11(2):203–28.

Rykalin NN. Plasma engineering in metallurgy and inorganic materials technology. Pure Appl Chem. 1976;48(2):179–94.

Schmidt-Szałowski K, Górska A, Motek M. Plasma-catalytic conversion of methane by DBD and gliding discharges. J Adv Oxid Technol. 2006;9(2):215–9.

Steinbach BP. An electrothermal fluidized bed carbon particle plasma reactor for hazardous waste treament. University of Missouri-Columbia; 1996.

Steinbach PB, Manahan SE, Larsen DW. The chemical reduction of small inorganic gases in an electrothermal plasma reactor. Microchem J. 2003;75(3):223–31.

Takarada T, Tamura K, Takezawa H, Nakagawa N, Kato K. The effect of pretreatment in a fluidized bed upon diamond synthesis on particles by chemical vapour deposition. J Mater Sci. 1993;28(6):1545–50.

Tsukada M, Goto K, Yamamoto RH, Horio M. Metal powder granulation in a plasma-spouted/ fluidized bed. Powder Technol. 1995;82(3):347–53.

Ua-amnueychai W, Kodama S, Tanthapanichakoon W, Sekiguchi H. Preparation of zinc coated PMMA using solid precursor by gliding arc discharge. Chem Eng J. 2015;278:301–8.

Uglov AA, Gnedovets AG. Effect of particle charging on momentum and heat transfer from rarefied plasma flow. Plasma Chem Plasma P. 1991;11(2):251–67.

Vaidyanathan A, Mulholland J, Ryu J, Smith MS, Circeo LJ. Characterization of fuel gas products from the treatment of solid waste streams with a plasma arc torch. J Environ Manage. 2007;82(1):77–82.

Visser J. Van der Waals and other cohesive forces affecting powder fluidization. Powder Technol. 1989;58(1):1–10.

Waldie B. Review of recent work on the processing of powders in high-temperature plasmas Part II—particle dynamics, heat transfer, and mass transfer. Chem Eng. 1972;261:188–93.

Wu CN, Yan BH, Jin Y, Cheng Y. Modeling and simulation of chemically reacting flows in gas-solid catalytic and non-catalytic processes. Particuology. 2010;8(6):525–30.

Yang JS, Bao WR, Zhang YF, Xie KC. Engineering application study of producing acetylene through coal pyrolysis in plasma reactor. Chem Eng. 2006;34(6):52–5.

Ye QZ, Li J, Xie ZH. Analytical model of the breakdown mechanism in a two-phase mixture. J Phys D Appl Phys. 2004;37(24):3373.

Zhu CW, Zhao GY, Hlavacek V. A DC plasma-fluidized bed reactor for the production of calcium carbide. J Mater Sci. 1995;30(9):2412–9.

Chapter 3
Non-thermal Plasma Fluidized Bed

Abstract In this chapter, the non-thermal plasma fluidized bed is introduced in detail. The thermal plasma fluidized bed includes gliding arc discharge fluidized bed, dielectric barrier discharge plasma fluidized bed, corona discharge plasma fluidized bed. Moreover, this chapter also introduces the research progress and applications of various reactors, and points out the existing shortcomings of them.

Keyword Non-thermal plasma fluidized bed

3.1　Gliding Arc Discharge Fluidized Bed

In combination with gliding arc discharge, spouted bed condition can be achieved with a suitable fluid velocity. Commonly used in the field of particle coating, spouted bed is particularly known to be advantageous in providing a uniform coating throughout the bed within a relatively short period of time. This is due to the excellent condition of heat and mass transfer achieved by circulation within the bed (Steinbach et al. 2003). The non-uniformity of surface coverage may be attributed to frequent inter-particle collisions and weak adhesive strength between the coating layer and PMMA surface.

It was equipped with a conic reaction chamber with ceramic lining (Fig. 3.1) and two, vertical, diverging, knife-like electrodes. Gaseous reactants, introduced at the bottom, kept the particles of the catalyst moving in the inter-electrode space. Within the fluidized bed, the reaction can be distributed in the overall working space. Thus, by incorporating gliding arc in a fluidized bed, the discharge can be stabilized and spread its impact into reactor volume. So the hybrid system which combined the gliding discharge (GD) and mobile (spouted) bed of catalyst particles moved by the gas stream across the discharge zone may be effectively used for the methane non-oxidative coupling. Previous research shows that the coupling of plasma with a fluidized bed of particles offers a large range of original properties for the development of new chemical reactors. Previous research shows that the coupling of plasma with a fluidized bed of particles offers a large range of original properties for

C. Du et al., *Plasma Fluidized Bed*, Advanced Topics in Science and Technology in China, https://doi.org/10.1007/978-981-10-5819-6_3

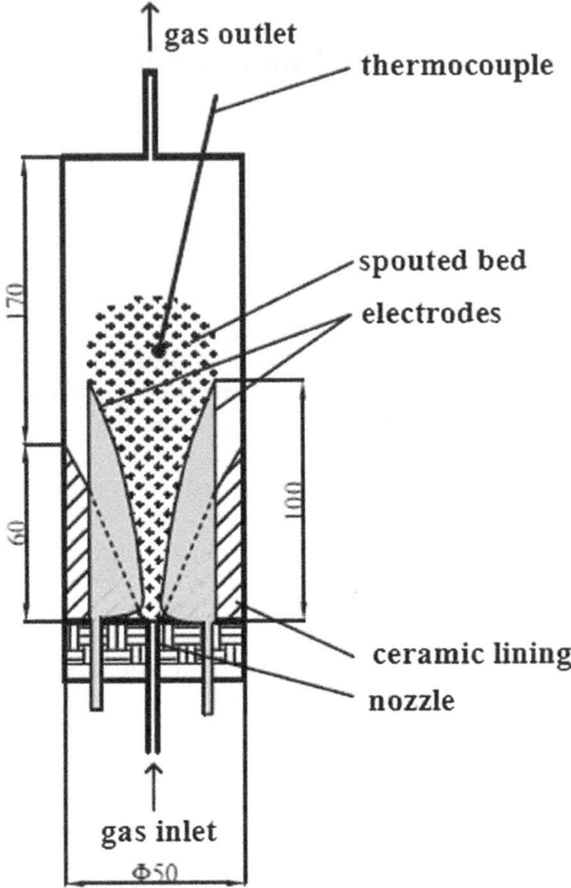

Fig. 3.1 GD reactor with a spouted bed of catalyst particles

the development of new chemical reactors. For a more uniform bed of very small particles, plasma discharge had to force its way through particles and its complete way of performing depends upon the balance of plasma power and fluidization forces. It was found that the presence of various catalysts especially Pt/Al_2O_3 leads to a higher conversion of CH_4 to C_2 and a much lower soot formation, which is generally ascribed to the interaction of the plasma and the catalytic process. On one hand, the active sites offered by catalyst surface changed the reformation mechanism of CH_4. On the other hand, active species generated in plasma can deposit on the catalytic surface and enhance the catalytic activity and the stability. Nevertheless, it should be mentioned that the presence of spouted bed of catalyst's particles in the GD zone may affect a number of process conditions, such as the electrical parameters of discharges, the hydrodynamics of gas flow across the discharge zone, temperatures of gas reactants, etc. To illustrate, this arrangement enabled the effective transfer of active species generated by plasma to the surfaces of catalyst particles. The circulation of particles in the spouted bed may be able to avoid the strong distortion of the discharge and local overheating of the catalyst in the stable bed (Savintsev 1990; Manieh et al. 1974; Schmidt-Szalowski et al. 2006; Flamant 1990; Lee and Sekiguchi 2011). However, the effect of the presence of the spouted bed on the discharge characteristic and the hydrodynamic characteristic has not yet been studied in detail. This arrangement enabled the effective transfer of active species generated by plasma to the surfaces of catalyst particles.

Previous research shows that the coupling of plasma with a fluidized bed of catalytic particles offers a large range of original properties for the development of new chemical reactors and has a potential to improve the selectivity of the plasma process. It is expected that this hybrid system could be used widely with commercial potential where an interaction of plasma process and catalytic reaction is needed, such as decomposition of VOCs, reforming of biogas and reduction of NO_X. The system will also be applicable for plasma-solids reactions including catalysts such as coal gasification.

However, a problem that cannot be ignored in the GD plasma fluidized bed is that radius of inlet nozzle is very small. So a severe issue must be taken into consideration that the particles can easily plug the nozzle. A proper designed distributor and well matched particles size and gas flow velocity are the keys to solve this problem.

3.2 Dielectric Barrier Discharge (DBD) Plasma Fluidized Bed

DBD plasma fluidize bed is a new technology related to plasma physics and chemistry. With the DBD plasma fluidized reactor, uniform and stable discharge is generated within the bed zone. The breakdown voltage within the fluidized bed is lower than that of the conventional DBD reactor. Also, it was found that the micro-discharge between the particles in the fluidized bed can significantly enhance the current density under the same condition, indicating that discharge in the DBD plasma fluidized bed is able to generate more active species. Here, the plasma is spatially confined to a scale comparable to the intrinsic dimension of a micro-discharge and therefore the total transferred charge is small. Therefore, the energy of the electrons and active particles in the plasma strengthen the reaction between the active particles and injected solid particles. Furthermore, inside the system, both the cycling water and the flowing gas can cool the bed, so the whole system can run continuously and the dielectric wall will not be broken down. The macro-temperature of the reactor is 315–320 K, so most of the organic materials will not be burnt, which enlarges its application area. As a consequence, this new method was seen as great opportunity for large scale application such as surface modification and catalytic process at low processing costs. Generally, DBD plasma fluidized bed can be divided as volume dielectric barrier discharge (VDBD) and

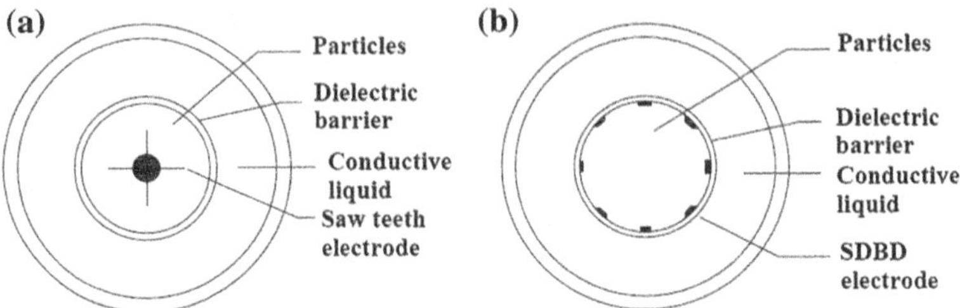

Fig. 3.2 **a** Cross sections of the volume dielectric barrier discharge reactor and **b** surface dielectric barrier discharge reactor

surface dielectric barrier discharge (SDBD) plasma fluidized bed, which are shown in Fig. 3.2 (Chen and Pfender 1983a; Liu et al. 1996).

VDBD has been proposed and utilized by Chen et al. (2009, 2006, 2008) for the surface modification of calcium carbonate powders using HMDSO in the He plasma. The powder surface energy was decreased greatly by coating an organic silicon polymer onto the powder surface. It was found that the technology is very promising for powder surface modification and pharmaceutical preparation because the particles are suspended in reactive conditions and can be coated homogeneously.

Moreover, investigation using such a reactor for seed treatment has been realized (Schmidt-Szałowski et al. 2006). Under the same growing environment, the abloom time of the treated coxcombs becomes longer than the untreated ones. Moreover, the flower size and the plant height of the treated coxcomb increase nearly two times, and the color of the leaves also turns from green-purple to purple. The exact mutating reasons are still unclear, and further gene analysis is needed. In our experiment, uniform discharge is an important factor for seed treatment and it can prevent the seeds from local burning.

In contrast to the VDBD reactor where plasma filaments (microdischarges) are formed in the gas volume between electrode teeth and dielectric barrier, in this case plasma formation occurs only on the dielectric barrier surface at the boundary of the electrode strips.

Another design for DBD plasma fluidized bed was described in previous studies. The novelty of the process was that an organosilicon precursor was introduced downstream of the discharge region. It was realized and elaborately investigated with respect to the efficiency of surface treatment in the afterglow region and the thermal exposure of the samples.

Scientific advancement towards short time processing of powders in the afterglow of a DBD is thus plausible considering the facts of better discharge control (no instability effects of particles on the discharge behavior), reduced particle charging by electron impacts, mild and homogeneous surface treatment (independence of DBD regime), and lower thermal loads. In principle, the basic concept of a PDR is also applied here for the efficient treatment of powders. In contrast to large mean free path of radicals at low pressure, the plasma is spatially confined at atmospheric pressure.

Additionally, the configuration of the PDR provides the opportunity to incorporate the particle plasma treatment in a continuous process chain. For this reason, the PDR system previously used by Spillmann et al. (2007) was adapted to meet the requirements of atmospheric particle treatment. However, large space requirements for the equipment and optical accessibility to the discharge device sometimes hinder their use.

Dielectric barrier discharge plasma has also been in combination with a circulating fluidized bed (Kroker et al. 2012). The effects on the hydrodynamic behavior have been investigated, and a proof-of-principle for coating copper oxide particles with a layer of silica will be given, showing that using a CFB with an incorporated volume dielectric barrier discharge reactor 20–30 μm CuO particles can be provided with a thin SiOx film. It has been shown that a certain minimum superficial gas velocity is needed to overcome the resistance created by the plasma; the plasma is shown to generate an additional pressure drop. It has been proven that for placing all electrodes at the column wall, a so-called surface dielectric barrier discharge, the deposition of powder on the electrodes is avoided and the cooling is more effective. Due to the fact that the SDBD plasma can be initiated at much lower voltage, the power input in the SDBD reactor can be better controlled using both voltage and repetition rate. Further investigations are needed to find out whether the SDBD plasma has a similar effect on pressure drop as observed when using the VDBD plasma. Overall, this fluidized bed plasma treating system is expected to provide improved coating rates and better temperature control of the coating process (Chen et al. 2014).

3.3 Corona Discharge Fluidized Bed

It is proposed that the future applications of corona plasma fluidized bed for plasma-assisted catalysis will most likely be in the treatment of gaseous waste and in the use of plasma to prepare and modify catalysts (Zhu et al. 2005). Actually, corona plasma technology has been utilized for soil remediation and many other plasma-catalyst hybrid techniques, which have shown satisfactory performance (Li et al. 2008; Nessim et al. 2009), for his allows for a significant synergistic effect, but the relatively high-pressure drop across the particle layer is problematic. A typical schematic of the corona plasma fluidized bed has been proposed by An (2004) since 1987, as shown in Fig. 3.3. A corona discharge fluidized bed consists of a gas injection system, discharge chamber and vacuum pumping system. Corona discharge plasma can be supplied with DC power with a wide range of power. However, investigation on corona plasma fluidized bed is still very limited yet. Features of such a reactor is also unknown, however, it can be expected that with corona plasma fluidized bed, high density of reactive species and low reaction

Fig. 3.3 Corona discharge
fluidized bed

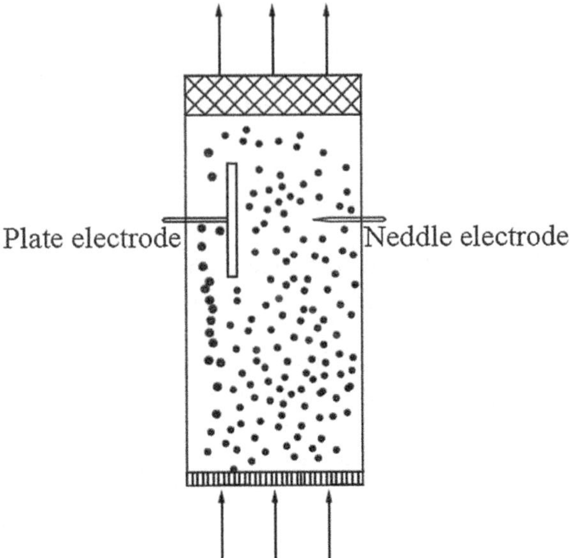

temperature and the uniform operation atmosphere can be achieved and therefore, excellent operation results can be gotten.

Much different from thermal plasma fluidized bed, non-thermal plasma fluidized bed shows totally different characteristic, and varied with various kind of plasma source. GD plasma fluidized bed is always taken as spouted bed. Temperature of GD plasma fluidized bed is in the range between thermal fluidized bed and typical low temperature plasma sources, therefore, GD plasma fluidized bed is more suitable for plasma-catalytic hybrid system. As for other non-thermal fluidized bed, the temperature is much lower and the active species is of great activity to offer extremely reactive chemical environment, which is of great importance for chemical reaction to improve the energy efficiency of the reactor.

References

An P. Research on discharge in two-phase mixture of gas and solid with its discharge characteristics in the nonuniform electric fields. Huazhong University of Science and Technology; 2004.

Chen G, Chen S, Feng W, Chen W, Yang SZ. Surface modification of the nanoparticles by an atmospheric room-temperature plasma fluidized bed. Appl Surf Sci. 2008;254(13):3915–20.

Chen G, Chen S, Zhou M, Feng W, Gu W, Yang S. Application of a novel atmospheric pressure plasma fluidized bed in the powder surface modification. J Phys D Appl Phys. 2006;39 (24):5211.

Chen G, Zhou M, Chen S, Lv G, Yao J. Nanolayer biofilm coated on magnetic nanoparticles by using a dielectric barrier discharge glow plasma fluidized bed for immobilizing an antimicrobial peptide. Nanotechnol. 2009;20(46):465706.

Chen X, Pfender E. Effect of the Knudsen number on heat transfer to a particle immersed into a thermal plasma. Plasma Chem Plasma P. 1983a;3(1):97–113.

Chen Z, Dai XJ, Magniez K, Lamb PR, Fox BL, Wang X. Improving the mechanical properties of multiwalled carbon nanotube/epoxy nanocomposites using polymerization in a stirring plasma system. Compos Part A-Appl S. 2014;56(56):172–80.

Flamant G. Hydrodynamics and heat transfer in a plasma spouted bed reactor. Plasma Chem Plasma P. 1990;10(1):71–85.

Kroker T, Kolb T, Schenk A, Krawczyk K, Młotek M, Gericke KH. Catalytic conversion of simulated biogas mixtures to synthesis gas in a fluidized bed reactor supported by a DBD. Plasma Chem Plasma P. 2012;32(3):565–82.

Lee H, Sekiguchi H. Plasma–catalytic hybrid system using spouted bed with a gliding arc discharge: CH4 reforming as a model reaction. J Phys D Appl Phys. 2011;44(27):274008.

Li MW, Gonzalez-Aguilar J, Fulcheri L. Synthesis of titania nanoparticles using a compact nonequilibrium plasma torch. Jpn J Appl Phys. 2008;47(9):7343–5.

Liu LX, Rudolph V, Litster J. A direct current, plasma fluidized bed reactor: its characteristics and application in diamond synthesis. Powder Technol. 1996;88(1):65–70.

Manieh AA, Scott DS, Spink DR. Electrothermal fluidized bed chlorination of zircon. Can J Chem Eng. 1974;52(4):507–14.

Nessim C, Boulos M, Kogelschatz U. In-flight coating of nanoparticles in atmospheric-pressure DBD torch plasmas. EUR Phys J-Appl Phys. 2009;47(2):22819.

Savintsev MI. Diffusion saturation in electrothermal fluidized bed. Met Sci Heat Treat. 1990;32 (11):842–5.

Schmidt-Szałowski K, Górska A, Motek M. Plasma–catalytic Conversion of Methane by DBD and Gliding Discharges. J Adv Oxid Technol. 2006;9(2):215–9.

Schmidt-Szalowski K, Krawczyck K, Mlotek M. Properties of a heterogeneous system of solid particles in gliding discharge plasma. In: 10th International symposium on high pressure low temperature plasma chemistry (Hakone X, 2006). 2006.

Spillmann A, Sonnenfeld A, Rohr PRV. Flowability modification of lactose powder by plasma enhanced chem vapor deposition. Plasma Process Polym. 2007;4(Supplement S1):S16–S20.

Steinbach PB, Manahan SE, Larsen DW. The chemical reduction of small inorganic gases in an electrothermal plasma reactor. Microchem J. 2003;75(3):223–31.

Zhu F, Zhang J, Yang Z, Guo Y, Li H, Zhang Y. The dispersion study of TiO2 nanoparticles surface modified through plasma polymerization. Physica E. 2005;27(4):457–61.

Chapter 4
Influencing Factors on Understanding Plasma Fluidized Bed

Abstract There is a comprehensive introduction about the influencing factors of plasma fluidized bed treatment effect in this chapter, which includes: the resident time, input power, gas flow rate, carrier gas composition, the design of the distributor, gas pressure, temperature, particle size and density, solid mass flow rate. All kinds of influencing factors are analyzed in detail, and the relevant precautions when using plasma fluidized bed are presented.

Keywords Influencing factors · Plasma fluidized bed

To attain expected performance of plasma fluidized bed, various influence factors including both plasma operation parameter and fluidization parameters can be used to moderate the processing of plasma fluidized bed.

4.1 Resident Time

In general, resident time of the particles in the plasma zone should be such that as much as possible interaction of plasma and particles is obtained at a given energy input.

In the plasma fluidized bed, the contact time or the reaction time between the plasma gas and the particles to be treated or the catalysts can influence the experimental results to a large extent. This can be generally attributed to the change in the inner plasma fluidized bed during the period. To illustrate, the gas-solid reaction can result in the generation of some active species or some intermediates which do not exist in the plasma reactor initially. These new formed species can change the chemical atmosphere in the plasma fluidized bed. Apart from the chemical atmosphere, a longer resident time can also influence the heat and mass transfer, the temperature of the particles and even the condense of the particles and the strengthening of the reverse reaction or the vice reaction was possible, which in turn affect its performance.

© Springer Nature Singapore Pte Ltd. and Zhejiang University Press 2018

C. Du et al., *Plasma Fluidized Bed*, Advanced Topics in Science and Technology in China, https://doi.org/10.1007/978-981-10-5819-6_4

To illustrate, Chen et al. (2006) used hexamethyldisiloxane (HMDSO) to modify the calcium carbonate in a He plasma fluidized bed, they found that with the increase of the resident time, the contact angle and the surface energy change of the treated powder changed significantly. The untreated powders were excellent hydrophilic, and then after 5 s retention time, the contact angle of water on the modified powders were 100°, and when the retention time reached 15 s, the water drop rolled freely on the powder surface. A full contraction and reaction between the plasma gas and the particles in the fluidized bed can achieve an excellent modification of the powders. Also, in a surface modification process, a too long resident time may lead to the condensation of the treated particles.

On the other hand, for some other reactions, like the cracking of the heavy hydrocarbon with some catalysts in a plasma fluidized bed, the resident time plays an important role in the selectivity and the efficiency of the system. In a study conducting methane and hexadecane co-cracking in an RF plasma fluidized bed, gas production rate has decreased as the catalyst operation time goes on. Inverse trend is seen for consumed power and process energy efficiency. About 35% efficiency reduction has occurred during 782 s catalyst operation time which it is attributed to catalyst activity decreasing caused by coke deposition generated with the processing of the cracking procedure. In this study, feed-catalyst contact time also affected the product distribution. While in another investigation of hydrogen production, H_2/CO selectivity was also found to be strongly dependent on the resident time (Vedrenne et al. 1991; Bromberg et al. 1998; Savinov et al. 1999; Cormier and Rusu 2001a, b).

Hence, choosing a proper resident time of the particles in the fluidized bed is very important for a better performance and a higher performance.

While in the circulating fluidized bed, the mean resident time of the injected particles in the plasma zone can be determined by varying the number of the circulations. Previous studies the control of the number of circulation can effectively change the resident time of the treated particles. Besides the change of mean resident time, circulation can offer overall contact with plasma gas and a more uniform treatment, especially for surface modification and therefore, sintering and overheating can be avoided at the same time.

4.2 Input Power

The nature of the plasma chemistry has outlined the dependency of the overall organic degradation rate on the energy input in a special range (Renzo and Maio 2007; Yan et al. 2013; Młotek et al. 2009). The effect of input power on the performance of the plasma fluidized bed lies in that the input power determines the amount of the energy input into the system provided for the ionization energy of electrons and then the energy needed for the radical generation. As a result, by increasing the power, both the electron temperature and electron density increase as noted in a study with a RF plasma fluidized bed (Nezu et al. 2003).

For the high temperature operation, the input power is related to the capacity to offer the heat needed. A higher input power means more heat can be supplied. However, it doesn't mean it can provide a higher energy yield.

Also, for other application of plasma fluidized bed, when the power is raised, both the electron and its density increase, and the reactive particles become more energetic. The applied power can increase the mean energy of the electrons, which leads to an increase at the rate of formation of chemically active species, such as OH, O, HO_2, H_2O_2 and O_3 etc. Besides, higher energy input favors more intense physical effect, which is greatly favorable for some chemical process, such as pollutant removal. The preliminary experimental tests reveal that the process stability can be increased by a higher RF power, lower pressure, lower gas flow rate and a lower amount of O_2 in the process gas (Kroker et al. 2012).

On the other hand, for non-thermal plasma fluidized bed, a higher density of the electrons with different energy induces the chain reaction and then the formation of various chemically active species. Hence, the reaction kinetics together with the concentration or even the types of the intermediates, result in the different product yields (Matsumoto et al. 1987; Taylor and Pirzada 1994).

In Thorsten Kroker et al' s study, it could be seen that the yield of H_2 and CO both increased with the power and the ratio of H_2/CO of increased at the same time (Kroker et al. 2012). The formation of the carbon black as function of power was much observable, when the power was below 40 W, no carbon black could be found while the power exceeded 40 W, the amount of the carbon black increased rapidly with power. The absence of carbon black can be explained by very stable carbon/oxygen bond.

However, a very different result was found in the study of conversion of CH_4 and CO_2 with a corona plasma fluidized bed (Chen and Pfender 1983). They found that an increasing of input power lead to the increase of the conversion rate, while the selectivity to H_2 decreased and the selectivity to CO increased. As a result, a higher power was followed by a higher ratio.

Change of the input power could also affect the feature of the product in surface modification. To illustrate, it was found that the oxygen/carbon ratio increased with a RF discharge power, which is caused by the increase of active species (Uglov and Gnedovets 1991). The contact angle decreases with the power, and the fiber becomes more hydrophilic.

On the other hand, a higher power might be not good, for too high a power sometimes leads to a two high temperature and the sinter of the product and energy consumption is another question, as Młotek found in his study, the unit energy consumption increased with an increasing input power (Młotek et al. 2009).

4.3 Gas Flow Rate

To save the gas consumption, the carrier gas of the fluidized bed also acts as the process gas of plasma. Therefore, flow rate of carrier gas is a vital parameter for the operation of a plasma fluidized bed due to the comprehensive mechanism. On one hand, the gas velocity can significantly influence the fluidization dynamic and the reaction time between the gas-solid. Also, discharge characteristic and therefore the stability and the production of the active species by plasma, all of which play an important role in the operation of a plasma fluidized bed (Arnauld et al. 1985).

With the gas velocity increasing, the particles in statics station gradually bubble and the intensity of fluidization of the particles became larger and larger. As the gas velocity reaches the original fluidized point, the particles are suspended in the upward flowing gas and the drag force between particles and the working gas counterbalances the weight of the particles. While the gas velocity excesses the transport velocity, the particles will be removed from the reactor (Liang et al. 2009; Heintze et al. 2003). For a better performance, a moderate gas-solid flow without large fluctuations is desired in the reaction zone. A proper gas-solid flow means sufficient interaction between plasma gas and solid phase and allows better heat transfer and mass transfer and avoids the possibility of the "dead corner", so they are very important in a particle or powder treatment (Schmidt-Szałowski et al. 2007), which means sufficient and uniform contact and of the particles and the active species and heat. Therefore, a homogeneous condition can be reached and also effectively lowers the possibility of local excessive heating part of agglomeration lowering the treatment effectiveness.

On the other hand, for the stable operation of a plasma fluidized bed, a stable discharge is required, therefore, a well developed gas-solid flow with moderate particle concentration but without large fluctuations is desired in the coating zone.

The gas velocity also affects the input power of the discharge zone (Vivien et al. 2002). For example, the discharge power of DBD plasma increased with the gas velocity and the gas production did the same way. However, when the gas flow rate is too large, the stability of the discharge would be destroyed (Du et al. 2014).

On the other hand, the increase of the gas velocity means the shortening of the resident time of the plasma gas in the discharge zone and a longer resident time can sustain full contact and reaction. To illustrate, when using a RF plasma fluidize bed to modify a kind of polymer powder with oxygen as carrier gas at given conditions, the oxygen functionality (such as C–O, C (O) O–) decreased with the oxygen velocity (Lee and Sekiguchi 2010). Also, plasma stability is not influenced by the plasma power, but depends strongly on the fluid dynamics of the two-phase flow system (Spillmann et al. 2007). Therefore, to generate stable plasma discharge, it is necessary to find the optimum operating conditions of solid circulation rate in the CFB reactor.

As a result, gas velocity could affect the operation of a plasma fluidized bed in many aspects, and a proper flow rate must be monitored for the best utilization.

4.4 Carrier Gas Composition

For a plasma fluidized bed, the processing gas includes the carrier gas, plasma gas and processing gas. To reduce the number of gas composition, the carrier gas of the fluidized bed can also act as plasma gas. In many plasma generator devices, gases such as Ar, He, H_2, N_2 and O_2 are commonly used as plasma working gas. Heat and momentum transfer of the plasma are influenced by the type and properties of the gas selected. The gas selected can be an inert carrier gas or take part in the chemical reaction. Furthermore, the gases used for plasma applications should be inexpensive and high good heating value. For different applications of plasma fluidize bed, different carrier gas can be used. To illustrate, argon may be the most frequently-used carrier gas in surface modification and many other applications. Argon is probably the most favored primary plasma gas and it is mixed with other secondary gases, because it is a monatomic molecule, having low ionization energy (15.8 V) and low chemical reactivity. Thus it can stabilize the plasma state and offer energetic electrons for the secondary gas to dissociate and produce the reactive radicals accounting for the reaction. While the breakdown of pure He was easily achieved with small input power, and the breakdown voltage is 1.03 kV at 5 W. The breakdown of pure Ar occurs about 2.5 kV and is followed by the arc like filamentary discharge. These values of breakdown voltage are similar to that of other atmospheric sources that are operated at ac or dc. Plasma discharges at near-atmospheric and higher pressure have a tendency to create plasma filaments which are followed by a rapid formation of arc or spark. The kind of discharge gas determines the stability of the glow discharge. Helium gives a stable homogeneous glow discharge, whereas nitrogen, oxygen, and argon easily cause the transition into a filamentary glow discharge. In the preliminary experiment, the effect of the contents (0–40%) of Ar in the He carrier gas on the plasma glow discharge was determined. With increasing content of Ar in the carrier gas, the breakdown voltage, the plasma glow volume, and the light intensity increase. In case of a 20% Ar mixture, the breakdown voltage is 1.6 kV at 75 W (Wang et al. 2011). For argon contents above 40%, the breakdown occurs at the input power higher than 200 W and has a tendency to create plasma filaments perpendicular to the electrode with increasing input power. Thus, a 20% Ar mixture for the stable glow discharge with economic considerations. The plasma power could be altered either by changing the current while fixing the plasma gas composition or by varying hydrogen concentration while fixing current (Chen et al. 2005). The overall effect of increasing power by either mechanism is increasing the plasma jet enthalpy and hence plasma jet temperature. It is expected that hydrogen addition affects the reaction rate by raising plasma power and enthalpy and not by involvement in the reaction. To verify this point, hydrogen concentration was kept constant at 33%, while power was increased by raising the current to equal the power of 45% hydrogen (Wang et al. 2009). It can be observed that for the same power input the rate of conversion was about the same for different hydrogen concentrations. This implies that hydrogen affects the rate only by raising the plasma power and plasma jet enthalpy.

Cold Plasmas treatments have been extensively used to modify the surface properties of materials without affecting their bulk properties. Surface modifications by grafting functional groups are necessary steps for most applications, in particular to improve adhesion. During the last ten years, the technology of Cold Remote Nitrogen plasma (CRNP) considerably extended and has demonstrated to be effective for polymer surface modifications. Surface chemical changes have already beers evidenced when exposing polymers to the C1tNp (Du 2014a, b). CRNP markedly differs from classical plasma as it is free of charged particles. Its reactive species are mainly atomic nitrogen, electronically and vibrationally excited N molecules with a long radiative or collisional relaxation time and vibrationally excited N molecules in the ground electronic state. The high density, the long lifetime of nitrogen plasma species and the low viscosity of the flow are characteristic of this kind of plasma. These properties allow realizing large volume chambers. Indeed, commercial reactors as large as 5 m^3 are now used to increase the surface free energy and so the hydrophilic character and the surface adhesiveness of polymers in too many industrial fields. Using a discharge oxygen plasma treatment, it was possible to increase the wet ability of a polyethylene (PE) powder. The contact angle with water, higher than 90° for the untreated PE, was equal to 51° after a treatment during 6 h (Ua-amnueychai et al. 2015).

The intensity of this emission is proportional to the square of the atomic nitrogen concentration. This concentration also depends on the dissociation rate of the nitrogen molecules in the discharge. This rate is very sensitive to the presence of impurities in the plasma gas. Particularly, an increase of nitrogen atoms concentration occurred if a small amount (<1%) of oxygen is added to the nitrogen flowing through the discharge. Indeed, the addition of oxygen to the plasma gas produces atomic oxygen O (3P). The O(3P) wall recombination, more efficient than the N(4S) one, explains the increase of nitrogen atoms concentration in these conditions.

Argon is a relatively cheap and widely used noble gas. The stable and long-lived ions stabilize the plasma and enlarge the electron density. Oxygen is commonly used as oxidant for any kind of metal oxide coatings. It forms negative ions and reduces the number of electrons in the plasma. If the solid surface is attached to the plasma, both of effects are possible; plasma is weakened by electron-surface recombination, but the emission of secondary electrons has a positive effect.

4.5 Design of Distributor

For a DC plasma torch fluidized bed, the design of the distributor is of vital importance for the stable and continual operation of the bed reactor to ensure efficient fluidization and avoid the possible overheating of reactor. Also the, the possible enlargement of the small holes on the distributor is another problem that needs considering (Matsumoto et al. 1987). For plasma fluidized bed, the distribution of the gas injected into the particle can be determined by the design of the distributor. Distributor with smaller pores can effectively control the gas flow and ensure a uniform transfer inside the fluidized bed. However, smaller pores may mean a demand for the vacuum system or the momentum cost. Bigger pores can lower the demand for the vacuum system but the quality of fluidization may be unsatisfactory. On the other hand, the shape and the pore number are also important detail of the design of the distributor. However, experiments in these aspects have been systematically studied yet.

4.6 Pressure

Pressure means a lot for the operation of the plasma fluidized bed. As mentioned in the previous section, plasma fluidized bed technology can be classified as low pressure and atmosphere plasma fluidized bed based on the operation pressure range kept inside discharge zone. Low pressure plasma fluidized beds have been widely utilized for various investigation and show very great potential. For low pressure plasma fluidized bed, the process pressure is a critical process parameter for operation (El-Naas et al. 1998b; Cormier and Rusu 2001a, b). On the other hand, this technique presents an important drawback at industrial levels as it requires low pressure to achieve the plasma state which usually means a higher cost but also that it increases the difficulty to obtain continuous processes. Compared with low pressure plasma fluidized bed, the main advantages are the elimination of vacuum systems, reduction of costs, the possibility to use continuous systems and treatment of materials which present high vapor pressure (Li et al. 2010). Therefore, it is very vital for the overall operation of the system to choose atmospheric plasma fluidized bed or a low temperature plasma fluidized bed for different applications.

Plasma parameter can be strongly affected by pressure. Furthermore, for low pressure plasma fluidized bed, the pressure range inside the reactor is also an important factor influencing the operation of the system. Under a constant input power, the change of pressure can significantly affect the discharge intensity, which can be described with the ion intensity of plasma. As found in previous studies, by enhancing the operating pressure, gas density increases and the mean free path of the gas molecules decreases and as a result the collision frequency of molecules increase, in turn suffers a large loss of the reactive species, which recombine to gas molecules (Uhm et al. 2014; Karches et al. 2004; Pajkic and Willert-Porada 2009). Hence, a higher can lead to a lower species concentration and the weaker impact of plasma gas with the particles.

A higher pressure can meanwhile increase the gas density inside and then reduce the average mean free path of the gas molecules. Therefore, the collision between molecules becomes more frequently the loss rates of the already activated species and the recombination reactions of different radicals to molecules are favored. As a result, with the increasing of the pressure inside the plasma fluidized bed, the plasma emission intensity and then ion density significantly decreased at a fixed position, especially for the position which is near the plasma source. Therefore, the utilization efficiency of input energy is also decreased.

What's more, at low pressure, heterogeneous surface reactions are favored while higher pressure is good for the homogeneous gas phase reactions, which make a difference in the gas-solid reactions. For example, in a process of nanoparticles formation, the nanoparticles could grow fast, its surfaces were coated less for the predominant gas phase reactions, the formed particles diameter distribution are shown as function of the process pressure ranging 6.0–21.4 mbar (Schmidt-Szalowski et al. 2006).

However, there existed a problem related to process pressure needing considering lies in the limited pumping capacity for a low pressure and with the decrease of the pressure sustained in the fluidized bed, the energy cost as well as the demand for the pump will increase to a large extent. Consequently, for different applications of plasma fluidized bed, the choice of the pressure of a low pressure plasma fluidized bed is very critical, and decision should be based on comprehensive consideration and optimization. To illustrate, a weaker impact of plasma treatment on the powder surface is expected at higher pressure, while for a higher intensity of surface modification, a lower pressure is in need.

At low pressures (1.5 mbar), heterogeneous surface reactions are prevailing and the formation of nanoparticles is inhibited. By increasing the pressure (2.0 mbar), the mean free path of the species in the plasma is reduced and homogeneous gas phase reactions and thus, nanoparticle formation is favored. Furthermore, the large error bar for 3.0 mbar indicates an inhomogeneous treatment of the substrate particles which can be attributed to the plasma property fluctuations which were observed visually at elevated pressures (Jung et al. 2004).

4.7 Temperature

For high temperature process with plasma fluidized bed, the effect of temperature on the behavior of a fluidized bed and spouted beds has been proved to be significant. The increase of temperature inside the fluidized bed increases the frequency of bubble formation and decreases bed viscosity, resulting in an increase in the fluidity of the bed. Also, the hydrodynamic of the fluidized bed and therefore the heat transfer and mass transfer can be influenced. The minimum spouting velocity increased with increasing temperature for large particles and decreased for small particles. For high temperature chemical process, such as the synthesis of calcium carbide the control of temperature based on thermodynamic equation is necessary for the highest production or the highest energy yield. Thus, according to the different treatment requirements of the operation, it is required to weigh the above two indexes.

As for other chemical process with plasma fluidized bed, it was found that a lower temperature tended to promote the generation of the active species such as HO radical and O_3 molecules due to the lower quenching rate. As for plasma-catalytic process, the effect of temperature would become complicated.

On one hand, catalytic process asks for a sufficient temperature for higher catalytic activity, high temperature may possibly cause sintering and catalyst inactivity (Savintsev 1990).

4.8 Particle Size and Density

In a fluidized bed, the property of the particles is doubtlessly playing an important role in determining fluidization regimes. As Geldart reported, the behavior of solids fluidized by gases fell into four distinguishable groups. The first group (A) includes materials having a small mean particle size and/or low density (less than 1.4 g/cm^3). The second group (B) contains most materials in the mean particle size and density range from 40 to 500 μm, from 4 to 1.4 g/cm^3. The third group (C) contains materials which lire cohesive. Powders of this type are extremely. The fourth group (D) includes material of large particle size and/or high density. The difficulties to fluidizing of these particles increase from A to D. Most materials contained in a plasma fluidized bed belong to the Group B. Particles belonging to groups A, B and D have been successfully used in fluidization systems by many authors as it will be presented next. However, particles in the C group have not been still present many problems and fluidization remains more like an art. However, some cases particles in the low limit of the micron range tend to agglomerate during fluidization helping the process to happen (Ye et al. 2004). Although it could give the impression that agglomeration would lead to the treatment of the whole agglomerate, this is a dynamic state between formation and disintegration of the agglomerates during the fluidization process (Zhu et al. 1995). Depending mainly on the forces between adjacent particles, the disintegration is governed by the shearing force due to bubble movement (El-Naas et al. 1998a). Therefore, mechanical vibration of the bed can be a very helpful option to solve this problem. The first of them, a down-stream reactor allows to treat considerable amounts of Carbon Black in Fluffy state (diameter below 10 μm). While for group D, particle with size larger than 1 mm has been successfully investigated in plasma spouted bed, which is featured by large, high velocity jet and internal solid circulation.

The particle size in the fluidized bed has a very large relation with the gas velocity needed to offer the fluidization, and therefore is related to the pumping energy. The mass transfer between the gas-solid also has a dependence on the chosen particle size. Fine particle has a large specific area for the solid-gas inter-action, and therefore enhanced mass and heat transfer can be achieved and a high efficiency could be gotten. For example, in the study of the synthesis of calcium carbide, it was found that under the same conditions, smaller particles with a mean diameter of 150 um exhibited a higher rate of reaction those of 600 μm.

However, the particle with a smaller size might ask for a higher demand for the distributor and filter trap. Similar to mass transfer, heat transfer is also related to the particle size in the plasma fluidized bed.

On the other hand, electrical resistance decreases with the increase of the particle size. For fine particles, the contact area diameter is less, and thus the electrical resistivity is more. Although the above expression is valid for equal sized particles in a cubic arrangement with the particles touching each other, it can also be extended to the case of a particulate fluidized bed. Because fine particles are easily separated by the fluidizing gas, the air gap adds to the specific resistance. Hence, a bed of fine particles always exhibits higher specific resistance. There seems to be no model that can be used to develop a single correlation to predict bed resistivity in terms of relevant parameters such as superficial gas velocity, particle size, current density, etc. The main reason for this shortcoming is the lack of experimental data obtained under identical conditions.

As a result, the particle size not only affects the fluidization dynamic, but also the electrical feature, in turn give birth to a profound effect on the performance of the overall reactor during the operation, the particle size may be changed due to inter-particle or particle-deflector collision, or simply to thermal shock of the particles as they are exposed to the plasma torch. Therefore, the effect caused by particle size on the overall performance is very comprehensive. Further investigation is still needed for full utilization of the reactor.

4.9 Solid Mass Flow Rate

The effect of solid holdup on the contact angle of the oxygen plasma treated High-density polyethylene (HDPE) powder (Liu et al. 1996). The stability of plasma can be negatively affected by interactions of particles and plasma. Particles may directly absorb RF energy or they could collide with excited gas species, namely reduce the number of electrons and its density. When solids concentration gets too high in the reactor, large parts of plasma volume can be eroded so that plasma could be extinguished (Shi et al. 2009). As can be seen, the contact angle increases with increasing solid holdup. At a very dilute solid holdup ($\varepsilon_s = 0.0003$), the HDPE particles might be under the influence of electrostatic force. After a few minutes operation, particles clung to the riser wall and grew up rapidly and then the riser was clogged with the HDPE clusters and consequent increase in system pressure to lead the plasma glow extinguished. However, this phenomenon was not observed with higher solid holdups greater than 0.001. When the solid holdup was 0.008, volumes of plasma glow were reduced apparently and contact angle of the treated HDPE powders changed slightly even with increasing the treatment time. Further increase in solid holdup in the riser, the plasma glow became unstable and results in the extinction.

The back-mixed fraction highly depends on the solid flow down rate and on the geometry of the exit. At very low solid mass fluxes, the solid phase passes the riser with concentrations less than 0.1%, which is not enough to form particle strands that are necessary for downward flow. This is the regime of pneumatic transport with plug flow reactor conditions. With increasing solid mass flux and higher solid concentration, back mixing dominates the solid flow. The internal recirculation gives rise to a higher particle concentration, enhances heat and mass transfer and smoothes the axial temperature profile. The back-streaming particles also change the velocity profile of the gas. In contrast to the empty tube, where the profile is flat with a thin boundary layer near the wall, the particles cause gas back mixing.

Therefore, the performance of plasma fluidized bed can be determined by various operation conditions, including the configuration parameters of the reactor as well as the operation parameters. Whereas, there exited other factors such as the electric field, the gas pressure, the ozone concentration, the electron density, and the active species, can also influence the performances. For example, because electron impact gives rise to the first step of the decomposition process and the generation of reactive species, the electron density is an important parameter in chemical reactions.

References

Arnauld P, Cavadias S, Amouroux J. The interaction of a fluidized bed with a thermal plasma: application to limestone decomposition. In: 7th International symposium on plasma chemistry (Eindhoven, 1985). 1985.

Bromberg L, Cohn D, Rabinovich A, O'Brie C, Hochgreb S. Plasma reforming of methane. Energ Fuel. 1998;12(1):1–18.

Chen X, Pfender E. Effect of the Knudsen number on heat transfer to a particle immersed into a thermal plasma. Plasma Chem Plasma P. 1983;3(1):97–113.

Chen GL, Fan SH, Li CL, Gu WC, Feng WR, Zhang GL, et al. A novel atmospheric pressure plasma fluidized bed and its application in mutation of plant seeds. Chinese Phys Lett. 2005;22(8):1980–3.

Chen G, Chen S, Zhou M, Feng W, Gu W, Yang S. Application of a novel atmospheric pressure plasma fluidized bed in the powder surface modification. J Phys D Appl Phys. 2006;39(24):5211.

Cormier JM, Rusu I. Syngas production via methane steam reforming with oxygen: plasma reactors versus chemical reactors. J Phys D Appl Phys. 2001a;34(18):2798.

Cormier JM, Rusu I. Syngas production via methane steam reforming with oxygen: plasma reactors versus chemical reactors. J Phys D Appl Phys. 2001b;34(34):2798.

Du CM. A plasma fluidized bed for the production of syngas from MSW. China patent 201410844203.9. 2014a.

Du CM. The treatment of VOCs by a plasma fluidized bed. China patent 201410849939.5. 2014b.

Du CM, Tang J, Mo J, Ma DY, Wang J, Wang K, et al. Decontamination of bacteria by gas-liquid gliding arc discharge: application to *Escherichia coli*. IEEE T Plasma Sci. 2014;42(9):2221–8.

El-Naas MH, Munz R, Ajersch F. Solid-phase synthesis of calcium carbide in a plasma reactor. Plasma Chem Plasma P. 1998a;18(3):409–27.

El-Naas MH, Munz RJ, Ajersch F. Modelling of a plasma reactor for the synthesis of calcium carbide. CANMetallQuart. 1998b;37(1):67–74.

Heintze M, Brüser V, Brandl W, Marginean G, Bubert H, Haiber S. Surface functionalisation of carbon nano-fibres in fluidised bed plasma. Surf Coat Tech. 2003;174–175(03):831–4.

Jung SH, Sang MP, Park SH, Sang DK. Surface modification of fine powders by atmospheric pressure plasma in a circulating fluidized bed reactor. Ind Eng Chem Res. 2004;43(18):5483–8.

Karches M, Takashima AH, Kanno Y. Development of a circulating fluidized-bed reactor for microwave-activated catalysis. Ind Eng Chem Res. 2004;43(26):8200–6.

Kroker T, Kolb T, Schenk A, Krawczyk K, Młotek M, Gericke KH. Catalytic conversion of simulated biogas mixtures to synthesis gas in a fluidized bed reactor supported by a DBD. Plasma Chem Plasma P. 2012;32(3):565–82.

Lee H, Sekiguchi H. Plasma-Catalytic hybrid system using spouted bed with gliding arc discharge. In: 240th American chemical society national meeting (Boston, 2010). 2010.

Li SD, Tian SH, Du CM, He C, Cen CP, Xiong Y. Vaseline-loaded expanded graphite as a new adsorbent for toluene. Chem Eng J. 2010;162(2):546–51.

Liang M, Chen J, Lin WM, Liu K. Thermodynamics study on solid phase decarbonization of high carbon ferromanganese powder. Ferro-Alloys. 2009.

Liu LX, Rudolph V, Litster JD. A direct current, plasma fluidized bed reactor: its characteristics and application in diamond synthesis. Powder Technol. 1996;88(1):65–70.

Matsumoto S, Hino M, Kobayashi T. Synthesis of diamond films in a RF induction thermal plasma. Appl Phys Lett. 1987;51(10):737–9.

Młotek M, Sentek J, Krawczyk K, Schmidt-Szałowski K. The hybrid plasma-catalytic process for non-oxidative methane coupling to ethylene and ethane. Appl Catal A-Gen. 2009;366(2):232–41.

Nezu A, Morishima T, Watanabe T. Thermal plasma treatment of waste ion-exchange resins doped with metals. Thin Solid Films. 2003;435(1):335–9.

Pajkic Z, Willert-Porada M. Atmospheric pressure microwave plasma fluidized bed CVD of AlN coatings. Surf Coat Tech. 2009;203(20):3168–72.

Renzo AD, Maio FPD. Homogeneous and bubbling fluidization regimes in DEM-CFD simulations: hydrodynamic stability of gas and liquid fluidized beds. Chem Eng Sci. 2007;62(1–2):116–30.

Savinov SY, Lee H, Song HK, Na BK. Decomposition of methane and carbon dioxide in a radio-frequency discharge. Ind Eng Chem Res. 1999;38(7):2540–7.

Savintsev MI. Diffusion saturation in electrothermal fluidized bed. Met Sci Heat Treat. 1990;32 (11):842–5.

Schmidt-Szalowski K, Krawczyck K, Mlotek M. Properties of a heterogeneous system of solid particles in gliding discharge plasma. In: 10th International symposium on high pressure low temperature plasma chemistry (Hakone X, 2006). 2006.

Schmidt-Szałowski K, Krawczyk K, Młotek M. Catalytic effects of metals on the conversion of methane in gliding discharges. Plasma Process Polym. 2007;4(7–8):728–36.

Shi TH, Wang ZC, Liu Y, Jia SG, Du CM. Removal of hexavalent chromium from aqueous solutions by D301, D314 and D354 anion-exchange resins. J Hazard Mater. 2009;161(2):900–6.

Spillmann A, Sonnenfeld A, Rohr PRV. Flowability modification of lactose powder by plasma enhanced chem vapor deposition. Plasma Process Polym. 2007;4(Supplement S1):S16–S20.

Taylor PR, Pirzada SA. Thermal plasma processing of materials: A review. Adv Perform Mater. 1994;1(1):35–50.

Ua-amnueychai W, Kodama S, Tanthapanichakoon W, Sekiguchi H. Preparation of zinc coated PMMA using solid precursor by gliding arc discharge. Chem Eng J. 2015;278:301–8.

Uglov AA, Gnedovets AG. Effect of particle charging on momentum and heat transfer from rarefied plasma flow. Plasma Chem Plasma P. 1991;11(2):251–67.

Uhm HS, Na YH, Hong YC, Yun J, Cho CH, Park YK. High-efficiency gasification of low-grade coal by microwave steam plasma. Energ Fuel. 2014;28(7):4402–8.

Vedrenne I, Herve T, Nikravech M, Amouroux J. C2 + hydrocarbons synthesis from methane in a plasma-spouted bed device. Stud Surf Sci Catal. 1991;61:207–12.

Vivien C, Wartelle C, Mutel B, Grimblot J. Surface property modification of a polyethylene powder by coupling fluidized bed and far cold remote nitrogen plasma technologies. Surf Interface Anal. 2002;34(1):575–9.

Wang Q, Cheng Y, Jin Y. Dry reforming of methane in an atmospheric pressure plasma fluidized bed with Ni/γ-Al$_2$O$_3$ catalyst. Catal Today. 2009;148(3):275–82.

Wang TC, Lu N, Li J, Wu Y. Plasma-TiO$_2$ catalytic method for high-efficiency remediation of p-nitrophenol contaminated soil in pulsed discharge. Environ Sci Technol. 2011;45(21):9301–7.

Yan B, Cheng Y, Jin Y. Cross-scale modeling and simulation of coal pyrolysis to acetylene in hydrogen plasma reactors. AIChE J. 2013;59(6):2119–33.

Ye QZ, Li J, Xie ZH. Analytical model of the breakdown mechanism in a two-phase mixture. J Phys D Appl Phys. 2004;37(24):3373.

Zhu CW, Zhao GY, Hlavacek V. A d.c. plasma-fluidized bed reactor for the production of calcium carbide. J Mater Sci. 1995;30(9):2412–9.

Chapter 5
Discharge Characteristic in the Plasma Fluidized Bed

Abstract In this chapter, the discharge characteristics of plasma fluidized bed are described in detail. Firstly, this chapter presents the factors that may affect the discharge of plasma fluidized bed. Then, the research status of discharge characteristics of plasma fluidized bed is analyzed, and the research on discharge characteristics is still lacking, so a lot of research is needed.

Keyword Discharge characteristic

Similar to dusty plasma, in plasma fluidized bed, the presence of particles would easily generate a unique discharge phenomenon due to the effect of charging particles. A particle in plasma gains an electric charge and responded to electric forces. The charge can range from zero to hundreds of thousands of electron charges, depending on the particle size and the plasma conditions. It arises from collecting electrons and ions from the plasma and sometimes from emitting electrons. In plasma in which emission processes are unimportant, the equilibrium charge is negative because the flux of electrons to uncharged surface is high relative to that of ions. Electrons can be emitted by the particle due to electron impact, UV exposure, thermionic emission and field emission. The first two are probably the most important for laboratory dusty plasma. Electron emission constitutes a positive current with respect to the particle, and if it is large enough, it can cause the particle to be positively charged.

On the other hand, when electron emission is significant, the equilibrium charge is positive, both of which can lead to the distortion of the original electric field (as shown in Fig. 5.1a and b) (Park and Sang 1999), and the most serious distortion of electric field occurred at the contact position between the charged particles, the degree of electric distortion is closely related to the capacity of the charge on the particles, the presence of the particles enhances significantly the electric field along the surface of the particles, when the particles are the glass bead with diameter of 0.5 mm, the enhanced electric field is 1.87 times as that without particles, while the electric field inside the particles is only 0.57 as the former electric field. When the particles are charged, the distortion would be much more complex.

© Springer Nature Singapore Pte Ltd. and Zhejiang University Press 2018

C. Du et al., *Plasma Fluidized Bed*, Advanced Topics in Science and Technology in China, https://doi.org/10.1007/978-981-10-5819-6_5

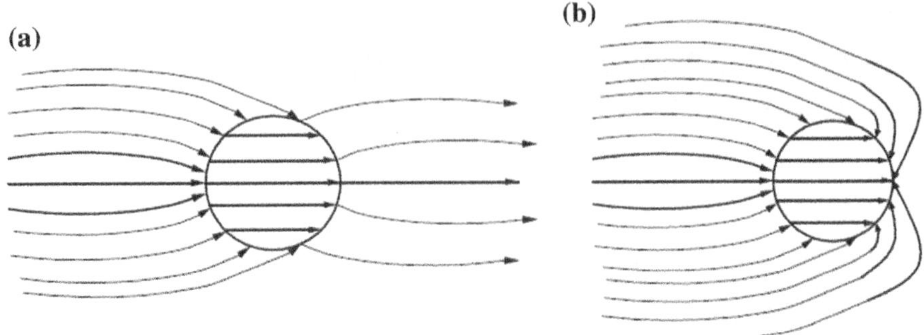

Fig. 5.1 The influence of **a** positively charged particles **b** negatively charged particles on electric field

The degree of electric field distortion is closely related to the electric capacity. Therefore, the determination of the electric capacity of the charged particles is an important index, which was found to be rarely a function of electric field position when in the uniform electric field. While taking the influence of particles between each other, the electric capacity can be described as M refers to the duty ratio. However, the equation is still unable to totally describe the electric field distortion caused by particles for the electric field generated by charged particles has not been taken into account. Also the frequent collision of particles as well as the complex discharge in fluidized bed can affect the electric capacity of the particles, and therefore, to accurately determine the electric capacity of particles as well as the electric distortion, much more experiments are in need.

Due to the electric field distortion and the microdischarge between the particles, discharge characteristics in a plasma fluidized bed are much different from that of conventional plasma techniques.

In the case of secondary emission, electron emission can be promoted by heating the electrons, perhaps by operating with a low gas pressure or using an electron-heating source such as microwave power. Far photo-emission, one could deliberately illuminate the plasma with a UV source. It may be useful to know that the particle's charge can fluctuate to a positive value even if it is not possible to charge it positively all of the time. Therefore, the charged particle in the discharge zone would unavoidably change the electric field inside a fluidized bed zone, also, electric force and electric field between charged particles would also significantly lead to the distortion of the overall electric field.

On the one hand, particles in the fluidized bed can act as the conducting media and decrease the discharge gap between the electrodes; on the other hand, negative charge on the particles can enhance the electric field near the negative electrode. Therefore, the break voltage of plasma fluidized bed is much lower than that of plasma chamber without particles. On the other hand, (Nezu et al. 2003) a dielectric barrier discharge has been described in which the discharge space is packed with glass spheres. Murphy and Morrow found that the presence of the glass spheres decreased the breakdown voltage. Note that the applied voltage in their experiment is ac, and the volume fraction of glass spheres is so high that edge breakdown will be produced. The reason is that the more macroparticles, the more they capture electrons, and then the higher is the breakdown voltage. It is worth noting that the effects of optical irradiation and electron emission from the macroparticle surface, which are influenced by the number of macroparticles and could strongly modify the breakdown process, have not been considered in the present model. On the other hand, we know that the effect of electron attachment of the electronegative gas will also result in an increase in the breakdown voltage. If the influence of the distorted field is not considered or $\varepsilon_i/\varepsilon_e = 1$, the case of electronegative gas is believed to be the limit to the TPM when the macroparticles become smaller and smaller. Ye et al. (2004) found that the change of pressure in the fluidized bed can also affect the breakdown voltage mechanism. Some studies have also reported that the larger the value of specific dielectric constant of the pellet, the lower was the initiation voltage of the partial discharge (Ye et al. 2004). Although the increase of the dielectric mismatch will result in an increase of distorted field, and consequently an increase of the average ionization coefficient, it will also result in an increase of the electron charge captured by the macroparticle, therefore, a decrease in the electron number density. In this case, the variation of breakdown voltage could not be simply determined.

For needle-pan discharge, as can be seen in Fig. 5.2, due to the enhancement of the electric field by negatively charged particles, the discharge intensity increased dramatically, and therefore the numerous uniform and stable microdischarge

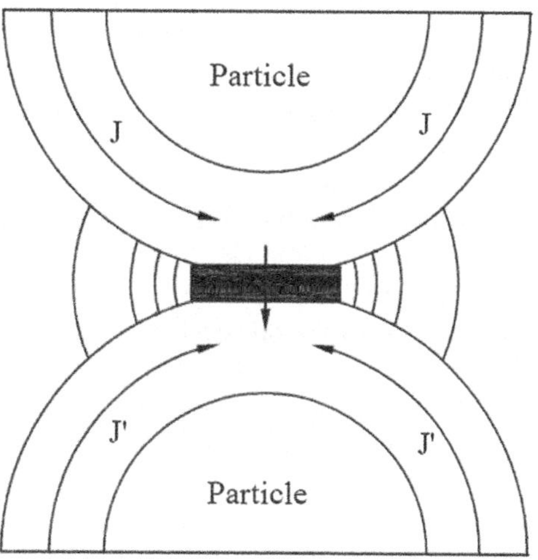

Fig. 5.2 Formation of microdischarge between particles

between particles can be generated. If the applied electric is intensive enough, the microdischarge can be lengthened till to the surface of reactor chamber, therefore, the discharge intensity can be enhanced, and numerous pulsed discharge can be generated and therefore the total current intensity can be increased to a large extent, which means that more active species (energetic electron) are produced, which is good for chemical reaction.

Researchers also found that in different plasma fluidized bed, discharge is much more uniform. The electrical parameters of the discharges were influenced by the presence of the spouted bed, and the saw-tooth shape of the voltage as well as current characteristics were most distinguished for the homogeneous system (with no spouted bed). When using a catalyst or alumina-ceramic particles, the voltage and current signals became smoother and the time period between the active parts of the discharge decreased due to the presence of the bed.

One of the possible reasons for these effects may be the combination of ions and electrons at the particle surfaces. The possible influence of electrically-charged particles on the breakdown voltage of the discharge should, however, be taken into account (Prat et al. 2000). For a relatively low gas velocity the particles are not homogeneously distributed: a local build-up of particles on the electrodes was observed. An increase of the superficial gas velocity to approximately 2.3 m/s appeared necessary to avoid the build-up of particles on the electrode (the space between electrode blades). The plasma is homogeneously distributed over the length of saw teeth electrodes and is stable for an extended period of about 4 h.

Actually, investigation upon overall discharge mechanism as well as discharge characteristics of various plasma fluidized bed and specific plasma fluidized bed is still very limited and falls behind the investigation of its application. However, this is very vital the scale up of such reactors and the right utilization. To illustrate, researchers found that the presence of micro-discharge between particles in the fluidized bed zone plays a positive role in $DeNO_x$ process, while for other operations, the presence of micro-discharge may sometimes leads to a worse performance for micro-discharge, lowering the overall plasma density inside the fluidized bed, also, for further research into the influence of particles on discharge characteristics, the charged quantity of the particle with difference features (size, shape, electric conductivity, etc.), the differences between plasma fluidized bed and dusty plasma, the specific discharge characteristics of specific plasma fluidized bed and the universal law of these reactors, the plasma component of the micro-discharge, etc. Nevertheless, the total mechanism of discharge in the TPM should include other effects, such as the field generated by the space charge, optical irradiation, electron and ion diffusion, electron emission from the surface of macroparticles, motion of charged macroparticles, deformation and coalescence of droplets (for liquids), etc.

References

Nezu A, Morishima T, Watanabe T. Thermal plasma treatment of waste ion-exchange resins doped with metals. Thin Solid Films. 2003;435(1):335–9.

Park SH, Sang DK. Functionalization of HDPE powder by CF_4 plasma surface treatment in a fluidized bed reactor. Korean J Chem Eng. 1999;16(6):731–6.

Prat R, Koh YJ, Babukutty Y, Kogoma M, Okazaki S, Kodama M. Polymer deposition using atmospheric pressure plasma glow (APG) discharge. Polymer. 2000;41(20):7355–60.

Ye QZ, Li J, Xie ZH. Analytical model of the breakdown mechanism in a two-phase mixture. J Phys D Appl Phys. 2004;37(24):3373.

Chapter 6
Hydrodynamics of Plasma Fluidized Bed

Abstract This chapter systematically reviews the hydrodynamics of plasma fluidized bed and its progress. Firstly, the hydrodynamics of plasma spouted bed is introduced from the aspects of minimum spouted velocity, spoutable height, pressure drop and particle attrition, and the relevant formulas are given. Finally, the hydrodynamics of plasma fluidized bed was analyzed.

Keywords Hydrodynamics · Plasma fluidized bed · Plasma spouted bed

Flow behaviors as well as hydrodynamic are very important characterizations of properties of plasma fluidized bed. It is expected that the presence of plasma can alter the fluidization process in the reactor compared with the conventional reactor, based on the ionized gas, thermal effect or the electromagnetic process (Wierenga and Morin 1989). The presence of the plasma discharge appears to help stabilize the fluidization behavior (Rogers and Morin 1991). In a DBD bubble fluid reactor, bubble formation and fluctuation exited in the bubble fluid bed are effectively inhabited and the gas flow passes through the bed by means of little gas flow forcing through the particles (Park and Sang 1999).

Fluidized bed is the typical technology of multiphase fluidization. Hydrodynamic investigation of fluidized bed has been one of the research focuses, for hydrodynamic plays an important role in defining the performance of fluidized bed and still constrains large scale industrial application of fluidized bed. Fluidized bed exhibits very complex hydrodynamics due to the nonlinear interactions between fluid and particle and their own individual movement tendencies. Many parameters involved in the design and operation of plasma fluidized bed are based on the hydrodynamic characteristics of the fluidized bed. Significant progress in the fluid dynamic investigation of conventional fluidized bed has been achieved (Kumar 2008; Nezu et al. 2003; Şen et al. 2012; Flamant 1990; Rogers and Morin 1991; Grovender et al. 2001; Mochizuki et al. 1993). Fluid dynamic of the gas-solid phases varies with different applications and different structure of fluidized bed. Investigation of hydrodynamic of conventional fluidized bed can be classified as follows:

© Springer Nature Singapore Pte Ltd. and Zhejiang University Press 2018

C. Du et al., *Plasma Fluidized Bed*, Advanced Topics in Science and Technology in China, https://doi.org/10.1007/978-981-10-5819-6_6

(1) Measurement and analysis of the character parameters, including the minimum fluidization velocity, pressure drop, the pressure distribution, particle size distribution or mass distribution, back mixing behavior of the particles, velocity distribution, temperature distribution, etc. These parameters play an important role to describe the fluid characteristic of gas-solid phases and understand the fluidization inside the fluidized bed and then in turn affect the overall performance.

(2) Developing or modifying the numerical models to simulate the fluidization state and interaction between gas and particles phases. However, due to the complexity of the gas–solid flow process, most of the models used are empirical models or semi-empirical models. With the rapid development of computers technology, computational fluid dynamics (CFD) are playing increasingly more important role in the model of multi-phase fluidization characteristic and promotes. Among the developed models, two modeling approaches are used most to investigate the fluid dynamic, the Euler-Lagrange (E-L) and the Euler-Euler (E-E) models. E-L model focus on the physics of the fluid-particles interaction, and provides the maximum number of details but has some difficulties in individual particle dynamics. However, one modified model of E-L model, Computational Particle Fluid Dynamics (CPFD) model, can simulate large-scale fluidized bed assuming a numerical particle by combining particles of the same properties, with comparatively less time. But, the information about each individual particle cannot be provided by the Euler-Lagrange CPFD model. The (EE) approach is numerically feasible for larger holdups; these models describe the motion for each phase in a macroscopic sense where the disperse phase as another continuum phases and separate governing equations are solved for each phase. Coupling is achieved through the inter-phase transfer coefficients. It is computationally less expensive compared to the Euler-Lagrange model and also provides reasonably good details about the fluidized bed.

Euler-Lagrange model provides the maximum number of details but computationally more expensive when compute individual particle dynamics (DEM/DPM). Whereas, the Computational Particle Fluid Dynamics (CPFD) model, is a modified Euler-Lagrange model which assumes a numerical particle by combining particles of same properties, can simulate large-scale fluidized bed gasifier with comparatively less time. But, the information about each individual particle cannot be provided by the Euler-Lagrange CPFD model.

The (EE) approach is numerically feasible for larger holdups; these models describe the motion for each phase in a macroscopic sense where the disperse phase as another continuum phases and separate governing equations is solved for each phase. Coupling is achieved through the inter-phase transfer coefficients. It is computationally less expensive compared to the Euler-Lagrange model and also provides reasonably good details about the fluidized bed.

From the present survey, it is observed that almost in all fluidized bed gasification models only the outlet gas composition compared with experiment. There are few measurements available for comparison with detailed model results. Therefore, effort is still required to validate the truly comprehensive fluidized bed models. Depending on the description of the hydrodynamics, the fluidized bed models may be classified as the two-phase flow model, the Euler-Euler model and the Euler-Lagrange model. Detailed descriptions of each of these models are presented below. Francke and Amouroux (1997) have applied laser Doppler anemometry (L.D.A.) which can tolerate the extreme environment in a plasma fluidized bed simultaneously measure the local density and velocity of particle distribution in a plasma jet fluidized bed with zeolite particles.

In the study, due to the velocity distribution, they found that the upward and downward motion of the particles forms a shearing region near the walls of the fluidized bed plasma reactor.

Based on the differences of the particle velocity and particle density, the particles distribution inside the reactor was distinguished into three different situations as the external annular location, the intermediate location and the center location.

And particle attrition might be a question. The interaction of the plasma jet with granular material is limited to the central part of bed and extent of the interaction is more important for coarse materials than fine materials (Du et al. 2008). However, different application and different structure of fluidized bed generally lead to different hydrodynamic characteristic, so does the plasma fluidized bed. In the presence of plasma, electronic field and magnetic field are expected to form and create electrons and ions, which can also produce attendant heating processes and affect significantly the flow characteristic. What's more, the electrical field and the ionized gas/particles have a considerable effect on the nature of and strength of the inter-particle forces and the collisions between the particles and particles or particles and the wall, and hence affect the expected pressure drop, then the point of minimum fluidization point, the gas velocity distribution and in turn alter the bed behavior. Some researchers have tried to find out the differences between plasma fluidized bed and conventional fluidized bed hydrodynamic characteristic.

Hassan Hamdi and Jerzy Jurewicz (Hamdi 2000) also found that with the increasing of the plasma and operation time, the particle attrition became more serious, which easily lead to the instability of the plasma spouted bed (Du et al. 2008).

6.1 Hydrodynamics of Plasma Spouted Bed

The hydrodynamics and stability of plasma spouted beds are studied experimentally in a reactor unit having about 2 kg bed capacity (Matsumoto et al. 1987). The influence of reactor geometry: base design, cone angle and torch nozzle exit diameter as well as solid size are investigated. The particulate solid used during the tests is alumina (corundum) with mean diameter ranging from 250 to 700 μm. Two cone angles are tested: 200 and 600. The plasma gas used for most of the experiments is a mixture Ar/N$_2$ with 20% N$_2$. Gas flow rates vary from 10 to 50 LPM.

The plasma power used during the experiments lies between 10 and 20 kW. Characterization of bed hydrodynamics and stability is carried out by measurement of the pressure drop across the bed and by visual observation of the bed behavior for each set of parameters tested.

6.1.1 Minimum Spouted Velocity

Similar to the conventional fluidized bed, the minimum fluidization velocity in the plasma fluidized bed has some relationship with the particle size (d_p). For a plasma argon reactor, the minimum fluidization velocity varied with 2.24 d_p.

The dependence of the minimum fluidization velocity in a classical fluidized bed on the bed temperature has been successfully predicted by a famous correlation proposed by Wen and Yu (Wen and Yu 1966). Therefore, one can be expected that the Umf decreased with the bed temperature. The experimental results in a previous study, however, showed a total different trend, which is for the above studies, it was apparent that the plasma fluidized bed had different characteristics than those associated with conventional fluidization. The hydrodynamics of such systems needs to be evaluated precisely using the exact bed temperature, which should be measured instead of predicted using the Ergun equation (Wierenga and Morin 1989; Sathiyamoorthy 2010).

6.1.2 Spoutable Height

Different from the standard spouted bed, the maximum spouted bed depth (H_m) for plasma spouted bed as defined by Flamant (1990), corresponds to the height of feed material beyond which the plasma jet extinguishes. The relation between the H_m and d_p is as follow:

$$H_m \propto d_p^{-3/4} \tag{6.1}$$

In a spouted bed, the correlations for predicting H_m have been proposed and modified by Wu et al. (Wu et al. 2010) as where refers to particle diameter and the sphericity. Various experimental results showed agreement with the equation mentioned above, and the relation between the H_m and d_p as follow:

$$U_{ms} = \left(\frac{d_p}{D_C}\right)\left(\frac{D_1}{D_C}\right)^{1/3}\sqrt{\frac{2gH(\rho_p - \rho_f)}{\rho_f}} \tag{6.2}$$

$$U_{ms}^f = \left(T_{fe}\right) \propto 1/3H_m^f \tag{6.3}$$

Here H is the spouted bed height.

Similar to the conventional spout bed, the maximum spouted height of the plasma spouted bed decreased with the particles diameter and the bed temperature (Du et al. 2015).

6.1.3 Pressure Drop

The relation of the pressure drop versus the mass flow rate indicated the minimum fluidization flow rate. For a plasma fluidized bed with a source of a high-frequency 2.45 GHz microwave plasma, in the presence of plasma, the pressure drop of the fluidized bed was much higher than that in the absence of plasma, which might be caused by both thermal and electromagnetic effects of plasma, and appears to help the stability of the plasma fluidized bed.

An iterative method for the calculation of average bed temperature by Ergun equation indicated that elevated temperature is also followed by the increasing of pressure drop. Wierenga and Morin (1989) have studied the variation of the pressure drop across the argon plasma fluidized bed but at a constant power density of 29.9 W/cm^3.

6.1.4 Particle Attrition

Special attention has to be paid to the situation of particle attrition in the fluidized bed. Particle attrition refers to the particle size reduction occurring in the plasma spouted bed, which would cause the hydrodynamic instabilities. Various experiments have been carrier out to investigate the influence of plasma operation on the particle attrition (Du et al. 2013; Lu et al. 2014; Ren et al. 2013). It was found that with the increase of the applied voltage and the duration time of plasma operation, particle attrition became more intense. Therefore, for a certain particle feature, proper duration time and plasma operation parameters are very important in respect to hydrodynamic stability.

The maximum spouted bed depth of plasma fluidized bed can also characterize the hydrodynamic of the reactor. The maximum bed depth in plasma spouted bed is much higher than that in the conventional spouted bed, as a result asking for a lower gas velocity to fluid the particle bed.

In plasma fluidized bed, it was found that the maximum bed height depends both on the plasma operating conditions and the bed temperature.

Numerical models to predict and describe the hydrodynamic characteristics in plasma fluidized bed in different applications have been proposed and modified.

In fact, to the author's knowledge, there exist no models yet dealing with the entire process, including the gas phase (plasma) physics (e.g. mass and heat transfer), plasma chemistry (electron-induced and heavy particle chemical reactions) and surface processes. However, some models have been reported that provide some insight into the possible synergistic effects of plasma catalysis, although the focus is mostly on the physical effects rather than the chemical effects.

The process for the tannery sludge with thermal plasma fluidized bed was modeled by Bai in order to offer a fundamental for the design, optimization and operation of the reactor. The hydrodynamic of the plasma-particle two-phase flow together with the interaction between the fluid parameter and the pyrolysis of the sludge were taken into consideration and played a vital role in numerical model proposed. The proposed model consisted of group of the description of the complex interaction and reaction between the plasma and treated particles, they are listed as follows: k-e double equation turbulence model to simulate the flow characteristics of the plasma phase, which is regarded in the turbulence mode, this model is consisted of the three equations (Ye et al. 2004; Rogers and Morin 1991),

$$\frac{\partial}{\partial t}(\rho\phi) + \frac{1}{r}\left[\frac{\partial}{\partial z}(r\rho u_z\phi) + \frac{\partial}{\partial r}(r\rho u_r\phi) + \frac{\partial}{\partial\theta}(r\rho u_\theta\phi)\right]$$
$$= \frac{1}{r}\left[\frac{\partial}{\partial z}\left(r\Gamma_\phi\frac{\partial\phi}{\partial z}\right) + \frac{\partial}{\partial r}\left(r\Gamma_\phi\frac{\partial\phi}{\partial r}\right) + \frac{\partial}{\partial\theta}\left(r\Gamma_\phi\frac{\partial\phi}{\partial\theta}\right)\right] + S_\phi \tag{6.4}$$

This model can offer insight into influence of the flow characteristic of the plasma-particle on the transfer of the substance and the resident time, to investigate the trajectory and the heat transfer of the different particles, to investigate the influence of the particle size and mass flow rate on the transfer of the particles in the plasma reactor. However, the simulation by this model is based on the ambient temperature two phase flow, and as a result this model is not sufficient to describe the plasma spouted bed (Rogers and Morin 1991).

$$(1 - \varphi)\rho_g v_g \cdot \frac{dv_g}{dz} = -\varphi\frac{18\eta}{d_p^2}\cdot v_{rel}$$
$$- (1 - \varphi)\rho_g \cdot g - (1 - \varphi)\cdot\frac{dp}{dz} \tag{6.5}$$

$$m_p\frac{du_p}{dt} = \sum F_p + m_p g \tag{6.6}$$

The major drawback of this model lies in the fact that, actually the chemical reactions considered are not sufficient to represent all complex situations and the effect of plasma source has not been investigated.

6.2 Analysis of Hydrodynamics of Plasma Fluidized Bed

To understand the characteristic of gliding arc plasma interacted with the effect of the fluidized bed, Fridmari et al. (1997) developed the model of interaction of gliding discharge and moving particles through the interplay of model of gliding arc plasma and hydrodynamic models of the fluidized bed, including the statistical model of bubbles coalescence and the model of fluidization condition. The researchers found that by incorporating gliding arc in a fluidized bed can help stabilize the discharge and the fluidization and the non-thermal plasma without strong thermal effect can effectively resolve the problem of particle attrition in the thermal plasma. The numerical model was based on the gliding arc parameters, fluidized bed parameters and the chemical reaction parameters and could act as a key to solve the problem during the operation, such as design principles, analysis of experimental results, and optimization of the performance (Fridmari et al. 1997).

The description of the fluidized bed state in the study was as listed below (Hu et al. 2015): I is an ionization energy, x (T_0) the thermal conductivity, T_0 the equilibrium gas temperature, this specific power will remain quasi-constant in the quasi-equilibrium zone. Equation 6.7 describes the power per unit length of the arc versus the thermal conductivity,

$$W = 16\pi \cdot (T_0)\left(T_0^2/I\right) \tag{6.7}$$

the self-inductance is presumed to be neglected. Thus, the ohm's law for the circuit including plasma channel, active resistor and power generator will be written as:

$$V_0 = RJ + Wl/J \tag{6.8}$$

Here V_0, R and J are respectively the open circuit voltage of the power supply, the serial resistance and current.

The arc current can be described as:

$$J = \left(V_0 + V_0^2 - 4WlR\right)^{0.5}/2R \tag{6.9}$$

It is suggested that during quasi-equilibrium period the current slightly decreases, the arc voltage is growing up as Wl/J, and the total arc power $P = Wl$ increases almost linearly with the length l. Further, consider when the arc length approaches the critical value:

$$l_* = V_0^2/(4WR) \tag{6.10}$$

The current falls to his minimal value

$$J_* = V_0^2/2R \tag{6.11}$$

But the voltage, electric field E and power approach respectively their maximum value when the arc length was equal to the critical value, voltage, the electric field and power respectively reached the maximum value, as the following Eq. 6.11. A description of the fluidized bed state, for fluidized spherical particles a resistance force and a buoyant force are balanced by particles mass:

$$C_d(R_e, \varepsilon)\pi d_p^2/4\rho_F/2(u^2/\varepsilon) + \pi d_p^3/6\rho_F g = \pi d_p^3/6\rho_s g \qquad (6.12)$$

The statistical model of bubbles coalescence gives a description of mechanism of bubbles initiation, movement and coalescence and thus the mechanism of particles circulation, distribution and mixing. The mentioned mechanism of the most important part of investigated phenomenon is impacted on arc-particles interaction and discharge effect distribution. By this model:

$$d_b = 0.853(1 + 0.272(u - u_{mf})^{1/3}(1 + 0.0684x)^{1.21} \qquad (6.13)$$

Here a resistance coefficient Cd for particles located in fluidized bed depends on Reynolds number:

$$\mathrm{Re} = u d_p/v \qquad (6.14)$$

Such a model of gliding arc plasma interacted with a fluidized bed could help to detect the coupled hydrodynamic characteristic in non-thermal plasma fluidized bed and also found help figure out the direction of such a reactor could be applied in, and as mentioned in the study, proper investigation of the hydrodynamic could achieve a better performance.

Plasma fluidized bed used to gasification was modeled by Lesinski et al. (1985) based on an Euler-Euler multiphase model. In the developed model, hydrodynamic characteristic was taken into consideration together with reaction models of drying, pyrolysis, homogeneous reactions and heterogeneous reactions. With this model, three groups of simulations were performed to study the influences of operating conditions. In the model, the Eulerian continuity equation is solved for each chemical species. Momentum and energy conservation equations are solved for each phase, the involved equations include the continuity Eq. 6.15, Momentum Eq. 6.16 and energy Eq. 6.17, as listed following:

Continuity equation

$$\frac{\partial}{\partial t}\left(\alpha_q \rho_q Y_i\right) + \nabla\left(\alpha_q \rho_q Y_i v_q\right) = m_i + S_i \qquad (6.15)$$

Momentum equation

$$\frac{\partial}{\partial t}\left(\alpha_q \rho_q Y_i\right) + \nabla\left(\alpha_q \rho_q Y_i v_q\right) = -\alpha_q \nabla p_q + \nabla \cdot \tau g + \alpha_q \rho_q g \qquad (6.16)$$

Energy equation

$$\frac{\partial}{\partial t}\left(\alpha_q \rho_q Y_i\right) + \nabla\left(\alpha_q \rho_q Y_i v_q\right) = -\frac{\partial p}{\partial t}\alpha_q + \nabla \cdot \tau g - \nabla q_p + S_p + Q_{pq} + m_{pq} h_{pq}$$

$$(6.17)$$

Numerical models to predict and describe the hydrodynamic characteristics in plasma fluidized bed in different application have been proposed and modified.

In fact, to the author's knowledge, there exist no models yet dealing with the entire process, including the gas phase (plasma) physics (e.g. mass and heat transfer), plasma chemistry (electron-induced and heavy particle chemical reactions) and surface processes. However, some models have been reported that provide some insight into the possible synergistic effects of plasma catalysis, although the focus is mostly on the physical effects rather than the chemical effects.

The general principle of the models was the same as the conventional fluidized bed, that is to say, the Eulerian approach and L-model.

Beat Urs Borer (Switzerland) proposed a model derivate from the work done by Durst to describe the accelerated gas/solid flow in the riser tube of the CFB coupled with a microwave plasma for SiO_x thin film deposition on particles. The developed model was based on Eulerian approach and composited of the momentum equations of solid phase and the gas phase combined with the continuity equations of the solid phase and the ideal gas law. In this model, the influence of the gas velocity on the overall fluid field, temperature distribution and particle concentration distribution in the reactor was investigated for an optimal velocity to achieve a high conversion rate with low cost.

The particle fraction as a function of superficial gas velocity for various gas pressures at RF power 5 W, where the laser scattered light intensity, I, is expected to be proportional to the particle fractions. The particle fraction at the center of the discharge tube increases with increasing superficial gas flow velocity; however, an unsteady mode is generated with as superficial gas velocity above 20 cm/s, where no significant influence has been observed with and without the existence of plasma. This unsteady mode may be generated from the change of the particle recirculation pattern through the oscillations (Du et al. 2010). Fluidization and plasma characteristics of medium pressure RF glow discharge plasma fluidized bed reactors are experimentally investigated and the following concluding remarks are obtained: (1) the oxidation of the powdered materials is successfully achieved by $Ar-O_2$ plasma; (2) the electron temperature decreases and the plasma density increases with fluidizing velocity; (3) the discharge configuration is almost uniform; (4) the discharge is stable under the majority of fluidizing conditions; (5) the particles are slightly positively charged by plasma; and (6) the oxidation of the powdered materials is twice as much and faster than the plasma fixed bed reactors.

The introduction of microwave-induced plasma into a bed of particles offers a unique environment. Depending on the operating mode that is set by the sliding short as well as by the probe depth, electric and magnetic fields are produced. Both fields create an electron and ion density profile that forms a corresponding temperature profile. The higher temperatures "hot spots" are usually found in the same region as the brightest plasma.

Data were taken with microwave-induced argon plasma operating in the mode for power densities ranging from 7 to 13 W/cm. For all measurements reported here, the downstream gas pressure, gas flow rate, and microwave power level were maintained constant, and measurements were recorded after the upstream pressure reached a stable value. For each configuration, the plasma "hot spots" were not located in the same regions. Frequently, the disturbances caused by fluidization and slugging of the bed caused the "hot spots" to shift position or be extinguished. This is significant, because the temperatures in the "hot spot" regions are greater than in the other regions of the bed. Pressure drop and mass flow rate measurements for 1.0 mm bead size illustrate both ideal and nonideal behavior. The gradual transition between the fixed and fluidized region was observed to be associated with a moving interface separating stationary and moving particles. This nonideal behavior appears to be more prevalent for the plasma data. As stated above, the location and intensity of the plasma "hot spots" were different for each run. This behavior resulted in varying minimum mass flow rates for beds operated under similar conditions. Plasma bed data were collected in both increasing flow rate order (forward) and decreasing flow rate order (reverse). Initially, plasma data were collected in the reverse direction to prevent bead melting. However, the data was taken in the reverse direction exhibit non-ideal behavior. As was done for non-plasma data, the measured pressured drops under plasma conditions were compared to the calculated pressure drop as a function of mass flow rate. The comparisons were made for runs at different downstream pressures ranging from 11 to 17 Torr. It can be seen from this figure that the data follow the same trend as non-plasma data, with small dependence on downstream pressure. However, the measured pressure drops are significantly higher than the calculated pressure drop, and this feature was characteristic of measurements from beds of smaller bead sizes as well. This behavior is unexpected, since the measured pressure drop should be close to the calculated value because no known time-average external force is present. This is, however, a clear indication that some additional force is acting to stabilize the bed contents against the onset of fluidization. This phenomenon might be due to the charge imbalance or buildup on the particles and the reactor wall, but no verification can be made at the present time. The magnitude of the additional force is several orders larger than would be expected from electrophoretic effects.

For DBD fluidized bed, Ma found that with the applied current, the bubble in the fluidized bed reactor can occur more easily, and that with the increasing of the applied current, the size of the bubbles decreases dramatically. For example, with a gas velocity of 0.3 m/s, applying an electric current of 8.4 mA can almost get rid of all the bubbles and form a much more stable and uniform particle bed. Also, with a larger applied current, the fluctuation in the fluidized bed can also be prohibited efficiently. Gas can pass through the particle bed in the form of gas jet with applied current instead of bubble, and therefore, the short circuit happening conventional fluidized bed can be avoided, resulting in enhanced transfer between fluid and particles. Therefore, plasma fluidized bed seems to be a good choice for gas–solid reactions, such as plasma-catalytic operation.

The results of feasibility demonstration indicate that the PFBR provides a flexible laboratory facility for the study of plasma–solid reactions. Argon plasma was created and maintained, using the empty cavity mode, in interstitial, intra-bubble, and intra slug regions of a fluidized bed. The vacuum system allows plasma-solid contacting in the PFBR to be conducted in a clean environment, with the solids, reactants, and products environmentally isolated.

Significant heating of bed solids is possible, as evidenced by the melting of the glass particles. An inferential method was suggested for the nonintrusive estimation of the gas temperature in a weakly ionized plasma. For mean pressures of 10–30 Torr, particle Reynolds numbers of 0.003–0.9, and Knudsen numbers of 0.005–0.2, it is apparent from the measurements taken in the absence of the plasma and the modified Darcy's law that the bed was operated in both the slip flow region and the laminar flow region. In the absence of a plasma, pressure drop and mass flow rate measurements through a bed of fixed and fluidized particles follow traditional high pressure results with ideal and non-ideal behavior occurring. In the absence of plasma, the pressure drop and mass flow rate at minimum fluidization can be calculated for a given mean pressure, particle diameter, and bed mass. In the presence of plasma, non-ideal behavior occurred more frequently due to an uneven temperature profile that is characteristic of a Microwave-induced discharge. In the presence of plasma, the measured pressure drop is greater than the calculated pressure drop required to raise the bed. There appears to be a stabilizing effect due to the presence of the high-frequency electromagnetic field and/or discharge (Wen and Yu 1966; Sathiyamoorthy 2010; Wu et al. 2010; Fridmari et al. 1997; Hu et al. 2015; Lesinski et al. 1985; Gerdes et al. 2006).

Modeling of plasma fluidized bed fluidization is an extremely important aspect for the further study and scale up of plasma fluidized bed, however. Ma has devoted a lot with respect to the study of the hydrodynamic and fluidization of plasma fluidized bed. He developed an E-E model to present the fluidization in the plasma fluidized bed (Fauchais and Vardelle 2000).

Taking electric force into consideration, the momentum equation of solid phase can be presented as

$$\frac{\partial}{\partial t}(\alpha_s \rho_s \overline{v}_s) + \nabla \cdot (\alpha_s \rho_s \overline{v}_s \overline{v}_s) = -\alpha_s \nabla p - \nabla p_s + \nabla \cdot \overline{\overline{\tau}}_s + K_{gs}(\overline{v_g} - \overline{v_s}) + \alpha_s \rho_s \overline{g} + \overline{F}_c$$

$$(6.18)$$

While for the fluidization

$$v_{mf} = \frac{1.08 \times 10^{-3}(Ga \cdot M_v)^{0.947} \mu}{d \rho_f} \qquad (6.19)$$

High-frequency, resonantly sustained plasma influence fluidization behavior through both thermal and electromagnetic effects. The effects of bed heating are most clearly seen in the fixed bed regime in the increased gas viscosity and pressure drops. The bed stabilization observed in the fluid bed regime may be related to the presence of the time harmonic electromagnetic fields and the accompanying polarization force.

References

Du C, Shi T, Sun Y, Zhuang X. Decolorization of acid orange 7 solution by gas-liquid gliding arc discharge plasma. J Hazard Mater. 2008;154(1–3):1192.

Du CM, Zhang LL, Wang J, Zhang CR, Li HX, Xiong Y. Degradation of acid orange 7 by gliding arc discharge plasma in combination with advanced fenton catalysis. Plasma Chem Plasma P. 2010;30(6):855–71.

Du CM, Huang DW, Li HX, Xiao MD, Wang K, Zhang L, et al. Adsorption of acid orange II from aqueous solution by plasma modified activated carbon fibers. Plasma Chem Plasma P. 2013;33 (1):65–82.

Du CM, Ma DY, Wu J, Lin YC, Xiao W, Ruan JJ, et al. Plasma-catalysis reforming for H_2 production from ethanol. Int J Hydrogen Energ. 2015;40(45):15398–410.

Fauchais P, Vardelle A. Pending problems in thermal plasmas and actual development. Plasma Phys Contr Fusion. 2000;42(12B):B365.

Flamant G. Hydrodynamics and heat transfer in a plasma spouted bed reactor. Plasma Chem Plasma P. 1990;10(1):71–85.

Francke E, Amouroux J. LDA simultaneous measurements of local density and velocity distribution of particles in plasma fluidized bed at atmospheric pressure. Plasma Chem Plasma P. 1997;17(4):433–52.

Fridmari A, Saveliev A, Nester S, Kerirzedy L. Nonequilibrium gliding arc in fluidized bed. In: 13th international symposium on plasma chemistry (1997, Beijing); 1997.

Gerdes T, Tap R, Bahke P, Willert-Porada M. CVD-Processes in Microwave Heated Fluidized Bed Reactors. Adv in Microwave Radio Freq Process. 2006;54–55(09):720–34.

Grovender EA, Cooney CL, Langer RS, Ameer GA. Modeling the mixing behavior of a novel fluidized extracorporeal immunoadsorber. Chem Eng Sci. 2001;56(18):5437–41.

Hamdi H. Contribution à la caractérisation du réacteur à lit soufflé par plasma [microforme]: application dans un procédé de gazéification du coke de pétrole. Université de Sherbrooke; 2000.

Hu MB, Dang SC, Ma Q, Xia WD. Stabilizing effect of plasma discharge on bubbling fluidized granular bed. Chin Phys B. 2015;24(7):288–92.

Kumar A, Dwivedi HK, Nehra V. Atmospheric non-thermal plasma sources. Int J Eng. 2008;2 (1):53–68.

Lesinski JMBJ, Meillot E, Debbagh-Nour G. Modelling of plasma entrianded bed coal gasifiers. In: 7th international symposium on plasma chemistry (Eindhoven, 1985); 1985.

Lu SY, Chen L, Huang QX, Yang LQ, Du CM, Li XD, et al. Decomposition of ammonia and hydrogen sulfide in simulated sludge drying waste gas by a novel non-thermal plasma. Chemosphere. 2014;117(117C):781.

Matsumoto S, Hino M, Kobayashi T. Synthesis of diamond films in a RF induction thermal plasma. Appl Phys Lett. 1987;51(10):737–9.

Mochizuki Y, Ono S, Teii S, Chang J. Fluidization and plasma characteristics of medium pressure RF glow discharge plasma fluidized bed reactors. Adv Powder Technol. 1993;4(3):159–67.

Nezu A, Morishima T, Watanabe T. Thermal plasma treatment of waste ion-exchange resins doped with metals. Thin Solid Films. 2003;435(1):335–9.

Park SH, Sang DK. Functionalization of HDPE powder by CF_4 plasma surface treatment in a fluidized bed reactor. Korean J Chem Eng. 1999;16(6):731–6.

Ren Y, Li XD, Yu L, Cheng K, Yan JH, Du CM. Degradation of PCDD/Fs in fly ash by vortex-shaped gliding arc plasma. Plasma Chem Plasma P. 2013;33(1):293–305.

Rogers T, Morin TJ. Slip flow in fixed and fluidized bed plasma reactors. Plasma Chem Plasma P. 1991;11(2):203–28.

Sathiyamoorthy S. Plasma spouted/fluidized bed for materials processing. J Phys Conf Ser. 2010;208:012120.

Şen Y, Bağcı U, Güleç HA, Mutlu M. Modification of food-contacting surfaces by plasma polymerization technique: reducing the biofouling of microorganisms on stainless steel surface. Food Bioprocess Tech. 2012;5(1):166–75.

Wen C, Yu Y. Mechanics of fluidization. Chem Eng Prog S Ser. 1966;62:100–11.

Wierenga CR, Morin TJ. Characterization of a fluidized-bed plasma reactor. AIChE J. 1989;35(9):1555–8.

Wu CN, Yan BH, Jin Y, Cheng Y. Modeling and simulation of chemically reacting flows in gas–solid catalytic and non-catalytic processes. Particuology. 2010;8(6):525–30.

Ye QZ, Li J, Xie ZH. Analytical model of the breakdown mechanism in a two-phase mixture. J Phys D Appl Phys. 2004;37(24):3373.

Chapter 7
Heat Transfer and Mass Transfer in the Plasma Fluidized Bed

Abstract This chapter introduces the heat transfer and mass transfer of plasma fluidized bed from three aspects: temperature distribution, heat transfer, circulation and mass transfer in the plasma fluidized bed. In terms of heat transfer, the effects of gas ionization effects, radiation effects and evaporation effects are introduced in detail, and the specific calculation formulas are given. In addition, this chapter summarizes the conditions and parameters of many studies and practical applications, which provide references for future research and application.

Keyword Heat transfer and mass transfer

Some of the characteristic of thermal plasma such as high temperature levels and high energy densities make the plasma very suitable for high temperature processes. However, the high viscosity of plasma makes the mixing the heat transfer with cold gases or particles difficult. The characteristics of a fluidized bed with its evolution with the temperature, the heat transfer fluidized bed-wall and plasma fluidized bed.

All these developments are now limited by the poor knowledge of basic properties of plasma fluidized bed and spouted bed, therefore, a presentation of heat transfer and mass transfer inside the plasma fluidized bed is in need for the further development of plasma fluidized bed.

7.1 Temperature Distribution in the Plasma Fluidized Bed

Another very important process parameter in a PFBR is the homogeneity of temperature. The temperature distribution of thermocouple is compared. The efficient dielectric heating of particles is visible from the temperature distribution developing upon microwave irradiation. Depending on the arrangement of the wave guides relative to the FBR-in-liner a hotter zone in the upper part of the microwave heated bed and a cold zone near the gas distributor is observed. In the case of the conventional heating a lower temperature at the top of the bed is measured, in

C. Du et al., *Plasma Fluidized Bed*, Advanced Topics in Science and Technology in China, https://doi.org/10.1007/978-981-10-5819-6_7

accordance with the heat loss to the fluidizing gas stream. The hot zone achieved by microwave heating along the height of the bed is beneficial for the CVD-process, because of a synergetic effect: the high particle velocity in the bed facilitates a high heat transfer, and the efficient heat transfer facilitates the pyrolysis, yielding a high reaction rate in this zone. Silicon-seeds which grow fast in this zone will move to the lower part of the FBR, where they will be continuously removed.

A strong decrease of temperature in the quartz in-liner and the typical temperature decrease in the thermal insulation are observed. The experimental data match with the calculation, with exception of the temperature within the thermal insulation. Because the alumina fiber insulation itself is heated by microwaves at higher temperatures, the temperature in the insulation is higher than the value calculated from thermal conductivity data provided by the supplier (Gerdes et al. 2006). Using resistance heating of the reactor walls, a 30 °C lower fluidized bed temperature is reached, however, it is necessary to overheat the in-liner by more than 100 °C as compared to the bed. In order to achieve this, the resistance heating elements are operated at their limit-temperature of about 800 °C. Comparison of conventional heating with microwave heating reveals an in-liner wall temperature of about 150 K less for microwave heating (Fauchais and Vardelle 2000).

7.2 Heat Transfer in the Plasma Fluidized Bed

Heat transfer in plasma fluidized bed and plasma spouted bed play a very crucial role, especially for thermal plasma fluidized bed. For reactions taking place in the thermal plasma fluidized bed such as coal/hydrocarbon cracking, plasma spraying, solid waste treatment as well as the preparation of ultra-thin film or nanoparticles (Shi et al. 2009a, b; Lee et al. 1985), the performance of the process is based on the effective interaction between the thermal plasma and the injected particles was dependent on the rate of the heat, mass and momentum transfer. To illustrate, in a plasma synthesis process, the complete fusion of the particles without evaporation is required for high quality coating (Rykalin 1976). Quenching rate by particles in immersed system is up to 50×10^6 K/s, which is much larger than the two-stage system (Flamant 1990). However, the net energy fraction transferred to the solid particle can be as low as 5%, which was ascribed to the relatively short resident time of the particle in the fluidized bed (Waldie 1972). To fully realize the potential of the thermal plasma fluidized bed/spouted bed, it is very crucial to further improve our insight in the physical and chemical mechanism of heat transfer between particles and the thermal plasma must be fully understood. Different from the conventional means for process intensification, we have made great efforts to further explore the special features of transport phenomena and chemical reactions in multiphase plasma reactors for both industrial needs and academic interests. Different from heat transfer in conventional fluidized bed, heat transfer between thermal plasma and particles is much more complicated due to the greatly different physical properties (specific heat capacity, viscosity and thermal conductivity) inside the plasma gas, gas ionization and recombination, the presence of the floating potential and local electromagnetic field, non-continuum situation etc. Over several decades, numerals surveys have been devoted to the heat transfer situation of the particles in such complex condition and different effects, including was also studied (Chen et al. 1995). These aspects include effects on heat transfer caused by variable properties due to the presence of large temperature differences between the plasma and the particle (Rykalin 1976), particle evaporation or sublimation (Chen et al. 1995), radiation (Chen and Pfender 1983a), and the Knudsen number (Uglov and Gnedovets 1991).

The heat transfer studies in the plasma-particle interaction have been focused on determining the effects of operating variables on the heat transfer flux. Therefore, a better understanding of the heat transfer between thermal plasma—particles is of great importance to find out optimal al parameters as well as the scale up of the reactor.

For the process of convective heat transfer with constant properties, the heat transfer coefficients can be derived from the semi-empirical correlations as follow (Rykalin 1976):

$$Nu = \frac{hD_P}{K_P} = 2.0 + 0.6 \mathrm{Re}^{1/2} \mathrm{Pr}^{1/3} \tag{7.1}$$

However, in plasma process, there existed a large temperature drop across the boundary layer, which in turn leads to the great gradient of specific heat capacity, viscosity and thermal conductivity of the flow between the plasma-particle inter-action. To determine the effect of variable properties on heat transfer, various modification of the heat transfer was proposed for the formula of Nusselt number (Du and Yan 2007; Du et al. 2008, 2014; Shi et al. 2009a, b). The results of different studies were in the similar trend, but there existed large discrepancies among different approaches and calling for further detailed investigations.

As mentioned above, there existed large temperature gradient across the boundary layer, which leads to the large variation of heat transfer to particles immersed in the plasma. Hence, the correction factors of different variation are caused by the plasma.

7.3   Gas Ionization Effects

Gas ionization is one of the most features in the thermal plasma compared with ordinary gas, and the presence of the electrons, ions and different neutrals in the plasma flow could result in the principle different characteristics of the heat transfer to the particles immersed in the thermal plasma. To account for the specific feature of ions, electrons in the gaseous phase on the heat transfer, numerous studies or models have been developed (Chen and Pfender 1983b; Chen and He 1986). The influence caused by the gas ionization may be due to the penetration of the electrostatic field a charged particle into the plasma but also that the presence of gas ionization can change the collision between the charged particles and the plasma flow. It was also found that the presence of ionization could enhance the specific heat flux over the surface. In the Chen and He's study, an analytical model accounting for the heat flux of specific species was given based on the velocity distribution function of gas particles and the flux of gas particles mentioned in (Chen and Pfender 1983b).

A systematic description of the heat transfer between the particles and thermal plasma was successfully proposed based on a kinetic theory approach (Bourdin et al. 1983) for the case of free molecules flow regime. The analytical descriptions are derived based on the following assumptions, which have in illustrated in various papers.

(1) The regime considered was the free molecule regime (Kundsen number is larger than 10).
(2) The Dedye length is negligible compared with the particle radius.
(3) The plasma consists of atoms, electrons and the singly ionized ions.
(4) Complete diffused reflection of gas particles takes place on the particle surface, while electrons contacting the particle surface are adsorbed into the surface. Ions recombination with electrons occurred with the combination with the release of the ionization energy.

The specific local heat flux caused by atoms was described as (Rykalin 1976):

$$
\begin{aligned}
q_a = a_a &\left[\int_{-\infty}^{\infty} \int_{-\infty}^{\infty} \int_{-\infty}^{\infty} \left(\tfrac{1}{2} m_a v^2\right) v_z f_a^- \, dv_x dv_y dv_z - \int_{0}^{\infty} \int_{-\infty}^{\infty} \int_{-\infty}^{\infty} \left(\tfrac{1}{2} m_a v^2\right) (v_z) f_a^+ \, dv_x dv_y dv_z \right] \\
= \frac{1}{4} a_a n_a \bar{v}_a k T_h &\left\{ \left[S_h^2 + 2\left(1 - \frac{T_w}{T_h}\right) \right] \exp\left(-S_h^2 \cos^2 \theta\right) \right. \\
&\left. + \sqrt{\pi} S_h \cos \theta \left(S_h^2 + \frac{5}{2} - 2\frac{T_w}{T_h} \right) [1 + erf(S_h \cos \theta)] \right\}
\end{aligned}
\tag{7.2}
$$

The specific local heat flux caused by electrons was described as (Rykalin 1976):

$$
\begin{aligned}
q_e = a_i &\int_{v_{ze}}^{\infty} \int_{-\infty}^{\infty} \int_{-\infty}^{\infty} \left(\frac{1}{2} m_e v^2\right) v_z f_e^- \, dv_x dv_y dv_z - \varphi_e(e\phi) + \varphi_e W_s \\
= \frac{1}{4} n_e \bar{v}_e k T_e &\left\{ \left(2 + S_e^2 + S_e \cos \theta \sqrt{\frac{E\phi}{kT_e}} + \frac{W_S}{kT_e} \right) \times \exp\left[-\left(S_e \cos \theta - \sqrt{\frac{e\phi}{kT_e}} \right)^2 \right] \right. \\
&\left. + \sqrt{\pi} S_h \cos \theta \left(\frac{5}{2} + S_e^2 - \frac{e\phi}{kT_e} + \frac{W_S}{kT_e} \right) \left[1 + erf\left(S_e \cos \theta - \sqrt{\frac{e\phi}{kT_e}} \right) \right] \right\}
\end{aligned}
\tag{7.3}
$$

Due to the fact that $m_e < m_a$, and thus $S_e < S_h \approx 1$, q_e (7.3) can be approximated as:

$$q_e == \frac{1}{4} n_e \bar{v}_e \exp\left(-\frac{e\phi}{kT_e}\right)(2kT_e + W_S) \tag{7.4}$$

The specific local heat flux caused by ions was described as (Rykalin 1976):

$$q_i = a_i \left\{ \frac{\int_0^\infty \int_{-\infty}^\infty \int_{-\infty}^\infty \left(\frac{1}{2}m_i v^2\right) v_z f_i^- dv_x dv_y dv_z + \varphi_i(e\phi)}{\int_0^\infty \int_{-\infty}^\infty \int_{-\infty}^\infty \left(\frac{1}{2}m_i v^2\right)(v_z) f_a^+ dv_x dv_y dv_z} \right\}$$

$$= \frac{1}{4} a_i n_i \bar{v}_i kT_h \left\{ \begin{array}{l} \left(2 + S_h^2 + \dfrac{E_I - W_S}{a_i kT_h} + \dfrac{e\phi}{kT_h} - 2\dfrac{T_w}{T_h}\right) \exp\left(-S_h^2 \cos^2 \theta\right) \\[2ex] + \left(\dfrac{5}{2} + S_h^2 + \dfrac{E_I - W_S}{a_i kT_h} + \dfrac{e\phi}{kT_h} - 2\dfrac{T_w}{T_h}\right) \sqrt{\pi} S_h \cos \theta [1 + erf(S_h \cos \theta)] \end{array} \right\} \tag{7.5}$$

Therefore, the specific heat flux caused by all the gas species can be expressed as:

$$Q = Q_a + Q_i + Q_e \tag{7.6}$$

In addition, the total heat flux to the whole particles was as follows:

$$Q_j = \int_0^{2\pi} \int_0^\pi q_j R_0^2 \sin \theta d\theta d\varphi \tag{7.7}$$

Therefore, heat flux caused atom, electron and ion can be explained respectively as:

$$\frac{Q_a}{4\pi R_0^2} = \frac{1}{8} a_a n_a \bar{v}_a kT_h \left\{ \left(S_h^2 + \frac{5}{2} - 2\frac{T_w}{T_h}\right) \exp(-S_h^2) + \frac{\sqrt{\pi}}{S_h}\left(S_h^4 + 3S_h^2 + \frac{3}{4} - \frac{T_w}{T_h} - 2S_h^2 \frac{T_w}{T_h}\right) erf(S_h) \right\} \tag{7.8}$$

$$\frac{Q_e}{4\pi R_0^2} = \frac{1}{8} n_e \bar{v}_e \exp\left(-\frac{e\psi}{kT_e}\right)(2kT_e + W_S) \tag{7.9}$$

$$\frac{Q_i}{4\pi R_0^2} = \frac{1}{8} a_i n_i \bar{v}_i kT_h \left\{ \begin{array}{l} \left[\sqrt{\pi}\left(\dfrac{1}{2S_h} + S_h\right) erf(S_h) + \exp(-S_h^2)\right] \times \\[2ex] \left(S_h^2 + \dfrac{5}{2} - 2\dfrac{T_w}{T_h} + \dfrac{E_I - W_S}{a_i kT_h} + \dfrac{e\phi}{kT_h}\right) - \dfrac{\sqrt{\pi}}{2S_h} erf(S_h) \end{array} \right\} \tag{7.10}$$

Based on the results calculated through the expression listed above, it was found that the effect of the q_e on the heat transfer density is more important than that of q_i distribution, and that the heat flux to the particles decreased with the increasing of the ratio of T_e to T_h when the T_e is constant. This may be due to the fact that ne and T_h decrease with increasing T_e/T_h, resulting in the decrease of the q_a, q_e and q_i.

7.4   Radiation Effects

Radiation loss form particles were negligible except for the temperature range of 2000–4000 K or in the case of low particle loading rates compared to other heat-transfer mechanisms (Chen and Pfender 1982). However, the effect caused by radiation will become more important for the following situations: (1) large particles; (2) high surface temperature and high emissivity; (3) low enthalpy difference between the surface of a particle and the plasma (Chen et al. 1985).

According to the numerical result in the previous study, the presence of radiation losses from particle surfaces will increase the net heat flux required for heating and evaporation. As a result, the boiling temperature of the particle is increased under the same conditions and in turn the heating process of the particle may also be changed. Thus, the radiation effect is very important for lower-enthalpy plasma (such as argon) and for materials with high boiling temperature (such as tungsten). The calculated results of the effects of radiation losses from particle ($r_s = 50$ μm) surfaces on the heating history, i.e. heating, melting, and evaporation period as function of the plasma temperature shown that the radiation effect was more important for tungsten particles immersed in argon plasma for the heating history of the particle was changed to a large scale. As for thermal plasma with higher enthalpies (Ar–H$_2$ or N$_2$), the effect of radiation on particle heating, melting, and evaporation is negligible except when the temperature difference between the plasma and the particle becomes too low or for large-diameter particles. The effect of radiation on heating history is more pronounced because of the higher surface temperature.

7.5   Evaporation Effects

A particle such as most of the metal particles and many nonmetal particles will undergo the g the evaporation (the surface temperature is approximately equal to the boiling temperature of the particle), which account for large amounts of mass transfer. Various studies were performed in terms of the effects of evaporation or sublimation on the heat transfer (Moissan and Dewar 1897). As mentioned in previous studies, the evaporation process could significantly affect the heat transfer by the following two ways. On one hand, different vapor was generated within the evaporation process of the particles, which could lead to the contamination of the plasma change the constitutions of the plasma. The electron number density increases significantly with the addition of, for example, copper to argon plasma at temperatures below 10^4 K. The low ionization potential of copper (7.7 eV) is responsible for this effect. This situation has a strong effect on electron properties, even for a small amount of copper vapor present in the argon plasma. The enthalpy, thermal conductivity, and to the product of density and interdiffusivity of argon and copper vapor varied with the copper vapor molar content to a certain degree.

Therefore, the heat transfer between the charged particles in plasma and particles would be enhanced. The calculated heat flux to a particle of 50 μm radius and $T_s = 1000$ K within the pure argon plasma shows the ratios of the heat fluxes to a particle immersed in a copper-argon mixture to that of pure argon plasma. It is obvious that Cu vapor in the plasma may have an appreciable effect on heat fluxes to a non-evaporating particle except for plasma temperatures above 15000 K, where the heat flux ratio approaches 1.0. With a mass fraction of copper vapor in the copper-argon plasma of 7.4%, the difference between the actual heat flux and values derived for pure argon plasma properties is as high as 41% at a plasma temperature around 9000 K. This difference reduces to 4% at 16,000 K. This variation is due to the nonlinear behavior of the thermal conductivity caused by minute quantities of Cu vapor (production of free electrons).

7.6 Solid Circulation and Mass Transfer in the Plasma Fluidized Bed

A multi-dimensional model of spouted bed was proposed recently by Du et al. (2014), and they propose velocities distribution for both the gas and the particles. The gas flows upward in the spout and in the annulus but the velocity in the former zone is one order of magnitude larger than that in the latter. Gas recirculation exists in the lower part of annulus. The particles move upward in the spout and downward in the annulus. The velocity reverses at the interface between the spout and the annulus, and inside the fountain. A parabolic like profile is observed using optical fibers mounted for measuring the variation of transmitted light by. On the contrary in the annulus, the annulus in which the voidage is larger than in the annulus and smaller than in the spout. The variation of spout voidage along the axis was predicted which was in agreement with the previous experiments, and it varies from at the injection orifice to about at the bed surface.

Various studies were also performed to determine the effects of different injection locations, particle properties (particle size, density and shape) and injection velocities on the heat transfer. Also, different models were proposed to simulate particle-plasma interaction in different reactors. Almost most of the investigations focused on the heat transfer during interaction of the particle and the plasma, however, most of the investigations had not been verified through experimental data. In addition, the overall effects caused by combination of different factors have seldom been detected. Hence, there is still a longer way for the investigation of heat transfer between particle and plasma to assist the design and sale up of the plasma fluidized bed with a larger energy utilization rate. However, successful scale up of microwave heated CVD-FBR processes requires a thorough investigation of microwave specific parameters, like e.g., penetration depth of the radiation into the fluidized bed, plasma ignition at high microwave power levels due to pressure variations and electrostatic charging, and reliability of microwave coupling into the dusty environment of a fluidized bed reactor. No general solution can be provided, so an optimized solution should be developed for each particular process. Sensitivity to temperature dependent dielectric and discharge properties has to be fully implemented into the process control, which in order to arrive at a successful microwave heated FBR´s on an industrial scale.

References

Bourdin E, Fauchais P, Boulos M. Transient heat conduction under plasma conditions. Int J Heat Mass Tran. 1983;26(4):567–82.

Chen X, Chen J, Wang Y. Unsteady heating of metallic particles in a rarefied plasma. Plasma Chem Plasma P. 1995;15(2):199–219.

Chen X, Chyou YP, Lee YC, Pfender E. Heat transfer to a particle under plasma conditions with vapor contamination from the particle. Plasma Chem Plasma P. 1985;5(2):119–41.

Chen X, He P. Heat transfer from a rarefied plasma flow to a metallic or nonmetallic particle. Plasma Chem Plasma P. 1986;6(4):313–33.

Chen X, Pfender E. Effect of the Knudsen number on heat transfer to a particle immersed into a thermal plasma. Plasma Chem Plasma P. 1983a;3(1):97–113.

Chen X, Pfender E. Behavior of small particles in a thermal plasma flow. Plasma Chem Plasma P. 1983b;3(3):351–66.

Chen X, Pfender E. Unsteady heating and radiation effects of small particles in a thermal plasma. Plasma Chem Plasma P. 1982;2(3):293–316.

Du CM, Shi TH, Sun Y, Zhuang X. Decolorization of Acid Orange 7 solution by gas-liquid gliding arc discharge plasma. J Hazard Mater. 2008;154(1–3):1192.

Du CM, Tang J, Mo JM, Ma DY, Wang J, Wang K, Zeng Y. Decontamination of bacteria by gas-liquid gliding arc discharge: application to. IEEE T Plasma Sci. 2014;42(9):2221–8.

Du CM, Yan JH. Electrical and spectral characteristics of a hybrid gliding arc discharge in air-water. IEEE T Plasma Sci. 2007;35(6):1648–50.

Du CM. A plasma fluidized bed for gasification. China patent 201310203334.4; 2014.

Fauchais P, Vardelle A. Pending problems in thermal plasmas and actual development. Plasma Phys Contr F. 2000;42(12B):B365.

Flamant G. Hydrodynamics and heat transfer in a plasma spouted bed reactor. Plasma Chem Plasma P. 1990;10(1):71–85.

Gerdes T, Tap R, Bahke P, Willert-Porada M. CVD-processes in microwave heated fluidized bed reactors. Adv in Microwave Radio Freq Process. 2006;54–55(09):720–34.

Lee YC, Chyou YP, Pfender E. Particle dynamics and particle heat and mass transfer in thermal plasmas. Part II. Particle heat and mass transfer in thermal plasmas. Plasma Chem Plasma P. 1985;5(4):391–414.

Moissan H, Dewar J. Nouvelles expériences sur la liquéfaction de fluor. C R. 1897;125:505–11.

Rykalin NN. Plasma engineering in metallurgy and inorganic materials tcchnology. Pure Appl Chem. 1976;48(2):179–94.

Shi TH, Jia SG, Chen Y, Wen YH, Du CM, Guo HL, et al. Adsorption of Pb (II), Cr (III), Cu (II), Cd (II) and Ni (II) onto a vanadium mine tailing from aqueous solution. J Hazard Mater. 2009a;168(1–3):838.

Shi TH, Wang ZC, Liu Y, Jia SG, Du CM. Removal of hexavalent chromium from aqueous solutions by D301, D314 and D354 anion-exchange resins. J Hazard Mater. 2009b;161(2–3):900–6.

Uglov AA, Gnedovets AG. Effect of particle charging on momentum and heat transfer from rarefied plasma flow. Plasma Chem Plasma P. 1991;11(2):251–67.

Waldie B. Review of recent work on the processing of powders in high-temperature plasmas Part II—particle dynamics, heat transfer, and mass transfer. Chem Eng. 1972;261:188–93.

Chapter 8
Scientific and Industrial Application of Plasma Fluidized Bed

Abstract This chapter introduces the application of plasma fluidized bed in detail, including the following four fields: metallurgy process, coal gasification and pyrolysis, environmental protection and materials. The metallurgy process includes: metallurgy extraction, synthetic of calcium carbide, alloy granulation, etc. In the field of gasification/pyrolysis of coal, including gasification/pyrolysis of coal for acetylene, pyrolysis/gasification of coal to syngas, biomass pyrolysis/gasification, gasification/pyrolysis of biomass for syngas, gasification/pyrolysis of biomass for bio–oil, gasification of solid waste cracking of heavy hydrocarbon and reformation of biogas. In terms of environmental protection, there are applications of abatement of VOCs, control of NO_x, sterilization of food, plasma modified catalyst for water purification and solid waste treatment. In the field of materials, there are applications of surface activation and functionalization, plasma enhanced chemical vapor deposition (PECVD) and synthesis of nanoparticles. The application of plasma fluidized bed has been widely used, and it has a wide application prospect.

Keywords Application · Plasma fluidized bed

In the last several decades, interest in the application of combination of plasma process and fluidized bed reactor has increased significantly for various industrial applications. In recent reviews, special attention is paid to the future challenge of plasma fluidized bed utilized for industrial engineering.

But looking ahead, still many obstacles remain to be surmounted for further research and development in order to meet the industrial demand.

The introduction of plasma fluidized bed into industrial practice is not only a big challenge for interdisciplinary research at the interface between plasma physics and life sciences, but also an option to develop new therapeutic strategies for several of nowadays hard–to treat diseases such as chronic wounds or MRSA.

Up to now there are no generally accepted criteria according to which atmospheric–pressure plasma sources can be assessed as to their suitability for various industrial reactions so far. Consequently, general characterization such as hydrodynamic, heat transfer as well as discharge characterization in the fluidized bed

© Springer Nature Singapore Pte Ltd. and Zhejiang University Press 2018
C. Du et al., *Plasma Fluidized Bed*, Advanced Topics in Science and Technology in China, https://doi.org/10.1007/978-981-10-5819-6_8

should be systematically investigated and summarized for different plasma fluidized bed. As the first step to help to evaluate plasma sources for biomedical applications according to risk analysis,

A plasma–generating device in combination with the fluidized bed is still in the process of evolution. Because development of a reactor of this type needs the process of evolution to take into consideration the engineering aspects of the plasma and the fluidized bed reactor at high temperatures, and this area has recently drawn the attention of many high–temperature scientists and engineers.

8.1 Metallurgy Process

The use plasma in metallurgy and high temperature chemistry dates back to 1897, when H. Moissan classical experiments were published (Moissan and Dewar 1897). Over several decades, the utilization of plasma based technology in metals processing operations has grown rapidly in research and industry. Thermal plasma offers an excellent method of generating very high temperatures without combustion, and may offer several advantages over conventional combustion technology. Plasma jet technology is rather prospective for reduction of oxides with low sublimation temperature which allows the substance to transfer into gaseous phase during its short residence in high temperature plasma zone (Gomez et al. 2008; Savintsev 1990; Chen et al. 1995; Chen and Pfender 1982, 1983a, b; Uglov and Gnedovets 1991; Chen and He 1986; Bourdin et al. 1983; Chen et al. 1985). It is established that hydrogen plasma provides thermodynamic advantages for reduction of oxides because of the presence of atomic, ionic, and excited states that are energetically much more effective in reducing metal oxides compared to molecules hydrogen (Chen and He 1986). Furthermore, the kinetic barrier associated with hydrogen plasma reduction processes can be removed because of higher local temperatures generated due to plasma assisted reactions. But it also decreases atmospheric pollution (Chen and Pfender 1982). Typically, the activation energy of plasma assisted reactions is lower compared to the corresponding standalone reactions. It is shown that both thermal plasma as well as nonthermal cold plasma can be effectively used for reduction purpose. Oxide dissociation and dissociative reduction can also be effectively used due to generation of very high local oxide minerals provides a potential future option. Hydrogen plasma intensifies the process as compared to Argon and Helium plasma. High hydrogen conductivity increases the heat transfer from plasma to dispersed particles and low viscosity coupled with low density lead to longer acceleration time of particles and their residence time in a high temperature zone (Kogelschatz et al. 1997). The attention on extraction of refractory metals by hydrogen plasma grew when researchers (Sathiyamoorthy 2010) studied the reduction behaviour of WO_3, Fe_2O_3, Ta_2O_5, Al_2O_3, TiO_2 and ZrO_2 in an arc of helium plasma jet, as an extension of the work done by Gomez et al. (Gomez et al. 2008). These advantages are not only due to increased operating temperatures but also due to the more reactive species available in the plasma medium. However, the accurate role of the high temperature and the reactive species in plasma zone for metallurgy and other high–temperature has not yet been investigated and determined.

On the other hand, metallurgical processes usually require large and very high–temperature reactors, which can result in high capital. Therefore, for successful application of plasma in extractive metallurgical processes, technoeconomic studies on the process and a commercially viable plasma reactor should be conducted.

8.1.1 Metallurgy Extraction

Over several decades, the utilization of plasma based technology in metals processing operations has grown rapidly in research and industry. Thermal plasma offers an excellent method of generating very high temperatures without combustion, and may offer several advantages over conventional combustion technology.

Metallurgical processes usually require large and high–power reactors. The most important existing and potential application of plasma technology include: electric arc furnaces, vacuum arc melting, plasma arc melting, and plasma fluidized bed (Nezu et al. 2003; Taylor and Pirzada 1994; Laroussi 2009; Thorley et al. 1959; Goldberger and Oxley 1963; Zhu et al. 1995; Sabat et al. 2015; Currier and Blacic 2000). Among these reactors, all of them belongs to furnace or fixed except the last one. The plasma fixed bed and furnaces represent established technology and have been used widely in industry for many years. However, metallurgy extraction in electrical furnaces induces some problems: (1) in order to keep the furnaces at high temperature, heat losses are very serious, and therefore the thermal efficiency is very low; (2) high-temperature operation requires high quality material for the construction of the facilities. Furthermore, because the present production in conducted in furnaces, the transportation processes of heat and mass between precursors are relatively slow, and so the kinetic process requires a relatively long time. All three factors result in high capital for production. Plasma fluidized bed reactor, which is emerging as an improved multipurpose high–temperature reactor, can overcome these obstacles and obviously replace the conventional reactor. However, developmental work must be undertaken before it is used in huge plants such as those in the steel industry: 100–MW plasma reactor would be required for a modest steel production rate of 250,000 tons per year (El-Naas et al. 1998).

Titanium tetrachloride can also be decomposed in a similar manner to produce titanium metal directly. A fluidized bed apparatus for producing titanium metal had been described. In this apparatus, titanium metal powder and glass beads were fluidized by $TiCl_4$ and H_2 which had been preheated to 60 °C. A glow discharge was produced between the electrodes immersed in the bed of electrically conducting titanium particles. The flow of current was not direct between the electrodes, but the potential applied between the electrodes generated sparks, and glow discharge was established across the conducting titanium particles, which grew in size due to the deposition of the titanium metal formed when $TiCl_4$ was reduced by

H_2 gas. Any unreacted $TiCl_4$ or H_2 was purified and recycled. Air distributed at the base of the reactor would permit continuous fluidization of the particle. In this way, particles would circulate rapidly upwards through the spout without overheating while fresh cold particles would fall downwards in annulus and then radically inwards to the spout.

For application of metallurgy extraction, various plasma fluidized bed has been investigated, including DC plasma torch plasma fluidized bed, RF plasma fluidized bed and MW plasma fluidized bed. Faced with the lack of oil and natural gas, plasma based technology using electric power for metallurgy is very promising and as a highly efficient plasma technology, plasma fluidized bed is expected to play an important role. However, the use of plasma fluidized bed in extractive metallurgy is still in its initial stages of the development and the involved studies are still limited. The proposed plasma fluidized bed is still faced with some problems prohibiting the further development. Within the thermal plasma fluidized bed such as DC arc plasma fluidized bed, particle with a larger size tended to stay longer in the jet due to their larger terminal velocities. Therefore, continuous operation is hard due to the agglomeration of larger particles.

For earlier investigations, DC plasma torch fluidized beds were preferred for metallurgy extraction. A DC plasma torch plasma fluidized bed firstly had been utilized for sodium carbonate roasting of vanadium mineral concentrate to extract vanadium as sodium metavanadate. In the application of DC plasma fluidized bed for calcining concentrated vanadium ore to extract vanadium in the form sodium metavanadate, conversion rate attained is about 19% in 7.4 min treatment. It was found that with the treatment time increased, the conversion efficiency doesn't necessarily increase due to the instability of the reactor. The salt tended to the agglomerate under high temperature and stopped the continual operation. Hence, those ventures into the development of plasma fluidized beds are advised to consider raw material availability and marketability of the processed material. It has been established that plasma can be used on a commercial scale with a high level of efficiency in production. Another limitation lies in the low conversion efficiency of the metal oxide according to the previous investigation. Poor circulation at the base of and instability in the high temperature atmosphere is another problem, which is suggested to be solved by reactor modification and operation parameter modification. A very attractive application by means of plasma fluidized bed has been proposed by Currier et al. of Los Alamos National Laboratory in USA (Currier and Trkula 2000; Currier and Blacic 2000). They propose an extractive metallurgy and oxygen recovery process well–suited for resource utilization in space based on a plasma fluidized bed.

The fluidized bed is a downer bed. With this plasma fluidized bed, solid materials can be fed from the upper side of the bed can be transported by the high velocity gas through the plasma zone. Hydrogen is chosen as process gas to offer atomic hydrogen plasma and forming a highly reductive atmosphere. By this way, oxygen can be extracted from the metal oxide and the metal or other structural materials can be achieved. During space exploration, oxygen could be used directly for life support or for propulsion (e.g. combustion of methane or in ion acceleration (plasma) propulsion schemes). It could also be used as an oxygen source for fuel cell powered robots engaged in surface exploration. The processed solids would be reduced from the oxides towards the base metals and could (with further refining) be used as structural materials. We screened extractive chemistry in plasma fluidized beds using hydrogen–argon plasma or in a radio frequency–generated hydrogen plasma. The plasma was generated using a microwave applicator (2.45 GHz) coupled directly to a quartz tube (the tube passed through the waveguide). The bed was fitted with a port just above the bed which allowed gas samples to be withdrawn for mass spectral analysis. We have successfully produced water from several surrogates of interest. To firstly investigate the feasibility of the reactor, $FeTiO_3$ (ilmenite) was used as a lunar surrogate (Currier and Trkula 2000). With this surrogate, water production from the hydrogen–argon plasma fluidized bed was fairly constant over time and significant changes in crystal structure were observed. A more complicated magnesium silicate mineral (olivine) was studied as a Martian surrogate, water can also be formed in an extended resident time.

Our other experiments in the plasma fluidized bed process indicate a general capability to form water from an even wider variety of oxide minerals. However, only a preliminary screening of these chemistries has been conducted and no concerted effort has yet been made to optimize the global kinetics. In order to do so, additional topics must be addressed in order to produce a compact design for space-based applications.

8.1.2 Synthesis of Calcium Carbide

Calcium carbide is an important industrial commodity and has been widely used in metallurgy field such as the desulfurization of steel and cast iron. Calcium carbide is generally produced by reacting calcium oxide and carbon in large electric arc at 2400 K on an industrial scale. However, the synthesis of calcium carbide by plasma process is faced with low efficiency and demand large energy consumption and therefore brings many technical problems such as the demand of special material and the further granulation of the product. Therefore, El-Naas et al. (1998) have done a lot of work and have made some significant improvements to energy-saving type plasma fluidized bed to strengthen further the cost competitiveness of the calcium carbide produced, which is shown in Fig. 8.1.

A plasma spout fluid bed process has been studied as a replacement of the present process. In this new process, a plasma spout–fluid bed reactor is used instead of an electric arc furnace. The reaction between carbon and calcium oxide takes place in the solid phase at about 2150 K, resulting in lower energy consumption. The good mixing characteristic of the spout–fluid bed and the confinement of the high temperature plasma within the bed improve heat transfer between the reactants leading to more efficient heating. Also, the sensible heat of the products can be recovered to heat the reactants and hence lower the energy requirements.

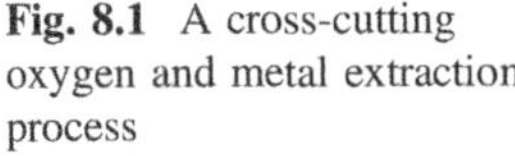

Fig. 8.1 A cross-cutting oxygen and metal extraction process

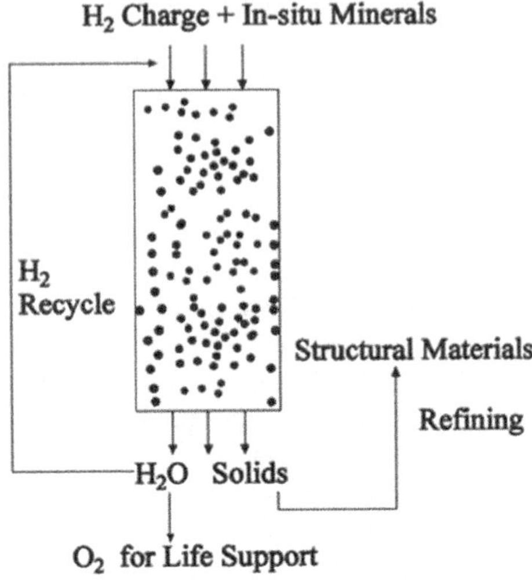

The energy consumption for a typical electric arc furnace is approximately 4 kWh/kg CaC_2. Production rate can be much higher and the product quality can be markedly better if the appropriate operations are used. Chemical equilibrium calculation conducted in their laboratory shows that the addition of argon or hydrogen gas in the plasma fluidized bed can effectively lower the reaction temperature to 1400–1500 °C. In plasma fluidized bed, it was found that temperature for the formation of CaC_2 decrease with gradually with increase of fraction of argon in the working gas. Working gas can not only offer heat needed for the reaction, but also remove away the gaseous by–products such as CO, CO_2 and then pushing the chemical reaction forward to reach equilibrium at the bed temperature under conditions without those gas by–products.

In many studies (Zhu et al. 2005; Li et al. 2008; Nessim et al. 2009; Park et al. 2005; Song et al. 2011; Vivien et al. 2002; Nikravech et al. 2015; Sachs and Wirth 2015; Bartolomeu et al. 2011; Arnauld et al. 1985), conversion rate up to 84.3% can be achieved at temperature of 1400–1450 °C within 40 min treatment. Therefore, energy needed for the synthesis of calcium carbide with a plasma fluidized bed is much lower up to 40% in the conventional electric arc furnaces. Also, the production process is very clean and there is very low material loss during the operation. Surface area and reactivity of the carbon source were found to be essential factors in determining of calcium carbide.

Later, study by El–Naas et al. found that chemical reaction is the controlling mechanism and mainly took place in the plasma jet zone (El–Naas et al. 1998). But the conversion rate is much lower, approximately 30%, which may be caused by instability of the bed due to the melting, sintering and agglomeration of the particles the difference may be ascribed to the different design and operation of the plasma fluidized bed. Therefore, the production rate can be much higher and the product quality can be markedly better expect for the well designed fluidized bed, and other problems must be solved.

Based on the investigations, a shrinking core, reaction model was also proposed. It is suggested that the position of the plasma torch should be modified to avoid melting of particles and bed instability. The same problem is also found in the application of DC plasma fluidized bed for calcining concentrated vanadium ore to extract vanadium in the form sodium metavanadate.

However, in order to scale up the process, it seems that many more tests should be performed, for instance, the dependence of the conversion rate on the reaction time, the effect of particle size of the carbon and the calcium oxide powders.

Reasons for the decrease of reaction temperature can be presented as follows: Argon with a certain flow rate not only initiates the plasma excitation and the increase the chemical activity of the reaction zone and then may be able to play the role of catalyst, but also blow away the generated gas products such as CO_x, which can also promote the continue reaction of CaO and C. On the other hand, the good mixing behavior between gas-solid and solid-solid phase makes sure the uniform temperature distribution and the enhanced mass transfer, and therefore much higher energy efficiency can be achieved. On the other hand, the good mixing ensure the produced caiculm carbied in the form of fine particles, therefore the energy consumption can be lowered compared with the convention furnace to make the deal with or lessen the aggregation and excess granulation problem.

8.1.3 Alloy Granulation

Granulation of metal powder has so far been limited and commonly used process uses water or organic solvent as binder. DC plasma fluidized bed has attracted interests in the field of grain production by offering extremely high temperature process without binder. One possible application would be alloy grain production for making metal catalysts such as Raney catalysts (Thorley et al. 1959).

Compared with the ordinary solidification and crushing process in respective of the production of the large alloy particles with metals or non–metals, plasma spout/ fluidized bed seems to be much more energy efficient due to the presence of the central high temperature zone. Furthermore, the solid circulation in the reactor shows good mixing behavior, as a result, the particles generated are fairly uniform in size and composition. Tsukada et al. (1995) in Japan used a dc plasma–spouted/ fluidized bed for granulation of the spherical alloy grains of 1–5 mm from iron powder and aluminum powder. They found that the grains exhibited a dense homogeneous core and porous non-homogeneous shell structure. It was also found that the addition of the seeds could increase the number of the large granular particles, which can be ascribing to the scavenging effect in the spout.

Plasma fluidized bed is chosen because of the flexibility of the adjustment of the operational parameters which seems to very valuable. Therefore, to utilize the plasma fluidized bed especially DC torch plasma fluidize bed in metallurgy extraction, there is still a long way to go. Another aspect is that the problem is damage of the electrodes by the materials containing oxygen and as well the sintering of the reactants on the electrode, therefore, the electrodeless reactor such as microwave plasma fluidized bed may be of great potential in this field.

8.2 Gasification/Pyrolysis of Coal

8.2.1 Gasification/Pyrolysis of Coal for Acetylene

Clean and efficient conversion and utilization of coal resource have been regarded as a key point to deal with the shortage of the oil resource globally and as an important way to ensure the safety of the world's energy structure. The characteristics of different thermal plasma torch systems have been studied extensively (Du 2014f; Attri et al. 2013). In a DC arc plasma, the temperature in the core of the plasma plume can be greater than 10,000 K, whereas in the marginal zones, the temperature decreases rapidly and the average operating temperature can be as high as 6000 K. In a RF plasma jet, the temperature at the central channel can reach up to 8000 K (Taylor and Pirzada 1994). A comparison of the main features of different plasma processes for waste treatment is given in. Over several decades, the production of acetylene with plasma gasification of coal has been an important aspect. As can be seen in Fig. 8.2, the gas phase of C_2H_2 pyrolysis has a complex chain reaction mechanism. Coal pyrolysis with one stage thermal plasma is

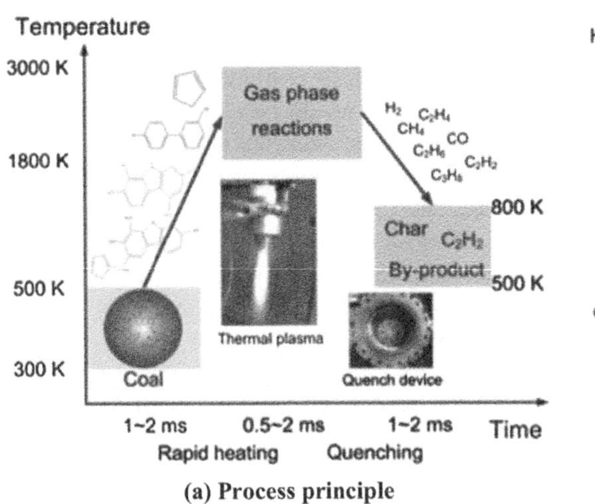

(a) Process principle

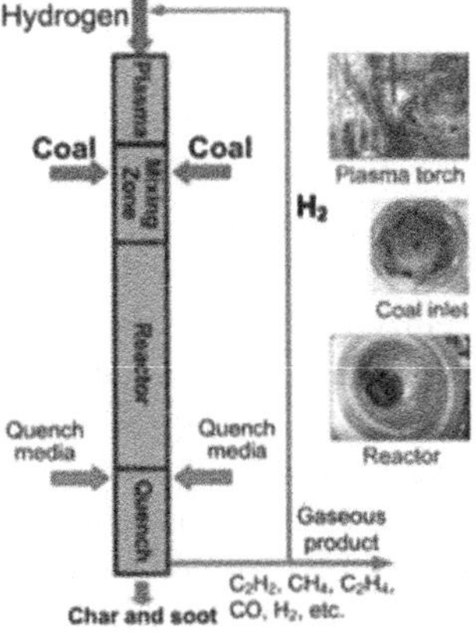

(b) schematic drawing of the plasma reactor

Fig. 8.2 Coal pyrolysis to acetylene in thermal plasma

characterized of simple operation, low water consumption, low carbon emission and low sulfur and NO_x emission (Hu et al. 2015). Therefore, plasma gasification offers a very promising way for coal chemistry. The most common plasma reactor used for coal gasification is downer reactor. Coal gasification inside the downer reactor can be divided into three stages, including initial devolatilization, gaseous reaction of the volatile and the quenching of the cracking products. The reaction time of coal gasification is very short, within the range of 10 ms. When injected into the reactor, the coal fine then mixes rapidly with the plasma jet in several milliseconds then instantly reach the thermodynamic equilibrium under the high temperature offered by plasma jet heating, leading to the formation of cracked gas with abundant acetylene. Compared to the calcium carbide method, one–step conversion from coal to acetylene in thermal plasma is a much cleaner coal utilization, which has no direct carbon dioxide emission, no request for a large amount of water and with hydrogen, carbon monoxide, methane, ethylene as byproducts. The principle of this process once the pulverized coal is heated to the temperatures above 1800 K within milliseconds, and a gaseous mixture will be released to the gas phase and soon convert to acetylene in the controlled temperature range of 1800–3000 K. However, acetylene will decompose to hydrogen and soot when the system temperature is lower than 1800 K. Therefore, there need a quench device at the downstream of the reaction chamber to efficiently prevent the decomposition of acetylene. In order to realize the above three important steps, the typical experimental setup consists of the plasma torch, the mixing section, the reaction chamber, the quench device and the separator.

The residual product can be mixed with other coal to be used as the feedstock for further gasification. For coal gasification, argon, helium, nitrogen, and hydrogen are usually used as processing gas. Previous study found that argon or argon-hydrogen plasma is best for the highest selectivity for acetylene, which is up to 95% (Lesinski et al. 1985).

A representative analysis of thermodynamic equilibrium of gas phase in coal pyrolysis process corresponds to an operation of the 5 MW plasma reactors in industry. Under the hydrogen atmosphere and high temperatures, the product gases of coal pyrolysis mainly consist of acetylene (7–10%), hydrogen (70–80%), carbon monoxide (10%), methane and little ethylene. The temperature before quenching changes from 1800 to 1500 K, and the mole fraction of acetylene in the gases drops from 8.31 to 2.76%. Therefore, the desired operation window of reaction temperature can be determined as 1800–3000 K. It can be summarized that the effective mass ratio of C/H in the gas phase and the quench temperature are the two dominant factors which play a dominant role in the yield of acetylene. Higher effective mass ratio of C/H would have a more positive effect on the reactor performance in terms of acetylene concentration. In general, the effective mass ratio of C/H in the gas released from coal is always less than 2.0. However, if alkanes or alkenes are chosen as the feedstock, the higher mass ratio of C/H (e.g., 3.0 for methane, 4.0 for ethane, 5.5 for decane, 6.0 for alkenes, etc.) will benefit the formation of acetylene in the product gas.

The company of AVCO has developed and industrialized a DC arc downer reactor for the generation of acetylene. High temperature up to 8000–10000 K atmosphere results in satisfactory performance, and the conversion rate of acetylene is as high as 33 and 67% of the coal can be effectively gasified. Previous studies have also found that the enthalpy of plasma is also a parameter for the production of the acetylene. A too high or too low value is bad for the production process (Lesinski et al. 1985). A too high enthalpy can directly lead to the formation of carbon black or a low value is not enough to initiate the gasification. Also, the particle size, heating value and other features of the coal fine, the operation pressure as well as design of the reactor can significantly influence the final product selectivity. To illustrate, when the power is fixed 6 kW/h, the acetylene yield was 22% (Chen and He 1986; Bourdin et al. 1983), and further decreasing the pressure can enhance the yield up to 28%. Alongside the length of the reactor tube, the concentration of acetylene is totally different. High volatility, high heating value, low oxygen content and good flow-ability are good for the highest yield of acetylene (Chen and Pfender 1982).

8.2.2 Gasification/Pyrolysis of Coal for Syngas

On the other hand, anther valuable product, syngas (i.e. the mixture of hydrogen and carbon monoxide) can be also produced by plasma gasification of coal, which can be date back to 1960s. It was found that adding steam or taking steam as plasma gas during coal gasification, the fraction of generated hydrogen and carbon monoxide can be increased, Geprgiev et al. found that the CO and H_2 were the main products is the off gas and the major mechanism of this process can be concluded as $C + H_2O = CO + H_2$. When steam is the only processing gas, the fraction of the generated syngas is up to 95–96%. Syngas generated by plasma gasification has great potential for application over a large field. It can be used for the ignition of the coal–fired boiler of large scale power plants to save a large amount of heavy oil. Also, it is found that the ratio of CO/H_2 can be controlled through the control of the operation parameter, such as the gas composition, according to various applications of the syngas. However, the research of this topic is still in the stage of basic research and there is still a long way to go. For a better gasification efficiency and easier operation, the physical and chemical feature and various operation parameters on the product composition and yield rate are needed to further investigate. Also, the thermodynamic equilibrium of the gasification reaction under high temperature should be determined and the macro kinetic model should be developed for the final optimization of reactor for high efficiency. A wide range of organic wastes treated by plasma process has been studied: MSW, used tires, paper mill waste (Spillmann et al. 2007), plastic waste, liquid and solid hazardous waste (Wang et al. 2010, 2011; Chen et al. 2005; Wang et al. 2009), Refuse Derived Fuel (RDF, i.e. mixture of plastics, paper, wood and dried organic material) (Wang et al. 2009; Schmidt–Szalowski et al. 2006; Flamant 1990), medical waste (Sarjeant and ROY

1967; Rohr and Borer 2007) and biomass wastes (Cormier and Rusu 2001; Ye et al. 2004; Zhu et al. 1995). On plasma technology must be added the nature of the plasma gas (Ar, N_2, H_2O, H_2, CO, CO_2, etc.), the specific enthalpy, the diffusion rate of plasma, the injected power, the thermal efficiency of the plasma torch and the technology of the plasma torch (DC, AC or RF) (Du 2014f; Taylor and Pirzada 1994; Bretagnol et al. 2004; Francke and Amouroux 1997; Uglov and Gnedovets 1991; Chen and He 1986; Bourdin et al. 1983; Chen et al. 1985; Moissan and Dewar 1897; Upadhya et al. 1986; Bullard and Lynch 1997a, b; Currier and Trkula 2000).

8.2.3 Gasification/Pyrolysis of Biomass

As a clean and renewable energy resource, biomass is the only one able to become the replacement of fossil energy to offer gaseous, liquid and solid fuel or other fundamental chemicals. Therefore, the development of proper technologies for the conversion and utilization of the biomass has become a very important topic. However, due to the low energy density, and low flow-ability and small density of the biomass, biomass in conventional fixed bed and downer bed is hard to fluidize, which makes it hard to realize efficient gasification economically. To enhance the efficiency of biomass gasification with thermal plasma, plasma spout– fluid bed has been used by various scientists (Santoianni et al. 2015). High contact efficiency and high circulation rate of in the spout–fluid bed can deal with the problem of low flow-ability of the biomass during the gasification to solve the coking and sintering problem of plasma gasification and therefore ensure the continuity and stability of the equipment. Different kinds of biomass have been studied in the plasma spout–fluid bed, including corn, rice and wood. Canola seed had been used as biomass source using a bench scale plasma spouted bed, fed with 150 g for one time. Argon was used as plasma gas with a mixing of a controlled amount of carbon dioxide (as reactive agent). H_2, CH_4, C_2H_2, CO and CO_2 were detected in the off–gas in the absence of tarry product. A maximum carbon conversion of 79% and a maximum oxygen conversion of 72% from the biomass feed to gaseous products have been obtained under our experimental conditions (Du et al. 2010). The product gas seems suitable for syngas applications. Considering the potential environmental benefits of biomass as an energy resource, biomass could be a more suitable feedstock to plasma pyrolysis processes than fossil fuels. The plasma spouted bed gasification of canola (colza) seeds is characterized by following advantages: (1) Process is very fast comparable with flash pyrolysis; (2) Absence of tarry by products at least at the process scale investigated. Very high overall energy efficiency up to 60% is comparable to the most efficient industrial pyrolytic gasification processes. It is expected, that for a continual operation of such reactor the efficiency will increase further high hydrogen and carbon monoxide content in the gas product accompanied only by light hydrocarbons: methane ethylene and acetylene, and the absence of sulphur, makes such production gas wry interesting as fuel for solid oxide fuel.

The possibility of employing medium temperature off–gas recycling into a plasma spouted bed gasifier is very interesting option to consider for achieving future balanced energy needs.

8.2.4 Gasification/Pyrolysis of Biomass for Syngas

Among various plasma pyrolysis of biomass, the liquefaction bio–oil belongs to forefront technology. Generated bio–oil offers an alternative of replacement of fossil oil. Investigation found that a plasma spout fluid bed can satisfy the demand for biomass pyrolysis to generate bio–oil. Quartz sand was placed in the fluidized bed acting as heating medium and once the biomass was fed into the reactor, intensive heat transfer from medium occurred, and the gasification took place. Biomass material and quartz sand were fed together in the plasma. For pyrolysis of biomass, the bed temperature is found to be important for the fact that a too high temperature would second cracking of the bio–oil to gaseous products and a too low temperature would lead to the un-complete reaction of the biomass.

8.2.5 Gasification/Pyrolysis of Biomass for Bio–Oil

With a pyrolysis temperature of 750 K, conversion rate up to 37.1% can be achieved (Santoianni et al. 2015). It has been found that the Acetaldehyde, hydroxyacetone hydroxy ethyl propionic acid ethylene glycol ester furfural acid can be detected (Currier 2000). However, due to relatively high oxygen content of the biomass treated, the heating value is very low, which is beyond the demand for practical application. The heat value of the bio–oil without any post–treatment is 18.0 kJ/g, while decarboned bio–oil is 21 kJ/g (Goldberger and Oxley 1963). After dewater, the heating value can be doubled. Therefore, to get bio–oil ready for application, the initial product has to be decarbon and dewater.

8.2.6 Gasification of Solid Waste

Plasma gasification of hazardous solid waste such as radiative wastes, chemical residuals, medical wastes and the fly ash of the incineration of municipal solid waste is most commonly investigated gasification of solid waste. High conversion rate can be achieved, but most of these processes are operated with an electrical furnace. Plasma gasification of municipal solid waste is able to recycle the syngas and other valuable chemicals. Therefore it is an efficient way for the final disposal of the organic solid waste (Du et al. 2014d; Currier and Blacic 2000). Plasma pyrolysis of organic solid waste almost doesn't produce any emission, and hazardous content such as sulfur and heavy metal can be fixed in the carbon black and therefore the gaseous products are very limited. It can be expected that when combining with a proper type of fluidized bed, plasma gasification of solid waste is a potential process. A plasma spout fluid bed may be able to be used for a special hazardous waste, sludge. It was found that with a spout fluid bed, the terrible flow ability and the sintering phenomenon can be avoided and keeping an ideal hydrodynamic in the reactor is very important for the plasma gasification of sludge (Sabat et al. 2015).

8.2.7 Cracking of Heavy Hydrocarbon

The purpose of hydrocarbons reforming is to allow the production of high added value products. Based on the feature of the processing gas, the reaction of hydrocarbon can be defined as oxidative and non–oxidative reformation. Plasma especially hydrogen plasma permits to obtain better yields as well as better selectivity towards the transformation (Thorley et al. 1959). Hydrogen plasma offers easily controlled temperature and highly reactive chemical atmosphere of H radical, which is of great importance for the cracking of hydrocarbon. However, given the cracking temperature of hydrocarbons, hydrogen plasma must be quenched to avoid the formation of coke and to favor the reforming reactions and then mixed with the hydrocarbon in a controlled moderate temperature. A fluidized bed seems to be adapted for the thermal control of the process and to ensure a high hydrogen concentration. On the basis of thermodynamic and kinetic considerations concerning hydrocarbon pyrolysis, a plasma–spouted bed device was developed (Yamamoto 1997; Şen et al. 2012). Cracking of paraffinic molecules such as CHI and $C_{16}H_3$ by Ar–H_2 plasma was carried out by continuous injection of hydrocarbon in the spouted bed. Control of temperature, residence time, and hydrogen concentration allow one to reach a cracking rate of about 95% without carbonization (Butscher et al. 2015). The role of radicals such as H and CH_3 on the conversion rate of heavy hydrocarbon was studied. Indeed, a number of transport properties of fluidized beds, close to those of plasma permit high energy transfer between plasma and solid particles. In the case of spouted bed in a parallelepiped reactor, the spouted particles create a "curtain" which separates the reactor into two thermal regions. On one side of the curtain, the quenching of plasma by particles generates a "hot region". The injection of methane in the hot region permits its dissociation in the spouted bed. Hydrogen is injected on the plasma gas in order to obtain its dissociation. The other side of the "curtain" is preserved from direct interaction of plasma. The temperature is moderate, homogeneous, and controlled by heavy hydrocarbon flow and by the plasma power. This region is adapted for cracking conditions of heavy hydrocarbons. When the cracking gas is utilized for special application having special demanding for light hydrocarbon such as C_4–C_8, catalytic cracking can be used. For production of C_4–C_8 molecules, the catalysis process in the bed is insured by zeolite catalysts which activity depends on their characteristics and on temperature. The spouted-bed is used for both of the plasma quenching and hydrocracking process catalysis. Indeed, the transport properties [viscosity (Du et al. 2008) thermal conductivity (Du et al. 2014d)] of the spouted-bed are of the same order of magnitude than the plasma ones, this allows an important transfer of heat and linear momentum between them. Consequently, there is a rapid quenching of the plasma giving a high radical concentration in the reactor. Also, the feature of the particles used for quenching of plasma permits to develop catalytic cracking and then to direct the reaction, to illustrate zeolites ZM760 was found capable to do so (Şen et al. 2012; Thorley et al. 1959; Butscher et al. 2015). A higher conversion rate up to 85% can be gained and the selectivity of the products can be modified. In order to increase the production of branched hydrocarbons, it is a general purpose to use acid catalysts.

Previous results show the presence of these radicals and their evolution in the plasma. In the case of a mordenite ZM760 bed, the increase of the temperature from 500 to 650 °C promotes the production of lighter hydrocarbons from the major product that is in C_4. The cracking rate is about 85%. The chemical properties of our inductively argon plasma are enhanced by hydrogen injection. Under similar temperature, cracking of n–hexadecane with Y–Na zeolite particles show a peak in C_2 (ethane + ethylene) while mordenite ZM 760 particles shows a peak in C_4 (isobutane). At high temperature (>650°C) the products distribution fats to a thermal one, whatever the particles nature. A maximum cracking rate of about 85% is obtained. In the case of a mordenite bed, the hydrogen of the plasma contributes to increase the light hydrocarbons (C_1–C_4) fraction. While using a Y–Na zeolite bed, it increases the more heavy hydrocarbons (C_5–C_8) (Butscher et al. 2015).

8.2.8 *Reformation of Biogas*

Methane, being the main component of natural gas (NG), is a valuable raw material for the synthesis of organic and inorganic products in the bulk–chemicals industry, as well as for the hydrogen–gas generation (such as in the petroleum refinery processes) the greenhouse gases methane and carbon dioxide have been used as alternative starting materials for the synthesis gas production. The plasma conversion of methane and other light hydrocarbons is regarded as one of the possible ways for making the utilization of natural gas resources more efficient from both economic and ecological points of view. The carbon dioxide (CO_2) reforming of methane (CH_4) to produce synthesis gas (H_2, CO) has received much attention for the utilization of both CO_2 and CH_4 resources, especially of the dilute methane resources (Schmidt–Szalowski et al. 2006; Currier and Blacic 2000; El–Naas et al. 1998; Lundholm et al. 2005). Several projects which are aimed at plasma application for methane processing have been proposed, for example: The moving particles in the plasma fluidized bed might bring new features to the synergetic effect of plasma and catalyst. In particular, the operation mode of a plasma fluidized bed would open new areas for plasma–assisted fluidization techniques. Catalytic particles were densely filled in the discharge gap (Attri et al. 2013; Renzo and Maio 2007). Total decomposition of hydrocarbons for obtaining hydrogen and carbon black both being valuable products, partial oxidation with steam, carbon dioxide, or oxygen into synthesis gas (mainly H_2 and CO), oxidative and non–oxidative coupling of methane for production of a various valuable chemicals: unsaturated hydrocarbons, aromatics, as well as conventional as well as plasma assisted reactors were used for this way of synthesis gas production (Arpagaus et al. 2005). Using biogas or biomass as starting material for synthesis gas production is an excellent idea because it relies on an exhaustless source of a renewable material (Du 2014a, b, f). The second big advantage is that biogas is very inexpensive. Last but not least getting rid of harmful greenhouse gases is another desired effect. Therefore, it is beneficial when the effect of these variations can be compensated for by changing the reactor conditions to obtain the desired product yield and selectivity. Furthermore, catalysts allow optimization of the product distribution. Unwanted side effects are carbon deposition on the surface of the catalyst and poisoning by sulfur. The typical source of sulfur is hydrogen sulfide, which is found in small amounts in biogas. The catalyst is deactivated and has to be regenerated. In a fluidized bed reactor, small amounts of the catalyst can be exchanged continuously without interruption of the reaction process. Both catalysts change the product composition significantly. Under homogenous conditions the maximum synthesis gas ratio (H_2/CO) was 1.3, whereas the highest ratio is found with the fluidized Cu/Al_2O_3 catalyst (up to 1.7). Interestingly the highest yield and selectivity of hydrogen is found with the fluidized Pd/Al_2O_3 catalyst (the synthesis gas ratio was up to 1.7). The catalyst temperature had an influence on the synthesis gas ratio as well, but not as significant as the plasma power. Carbon black production starts at 40 W following an exponential trend (at lower power no carbon black is observed), and amounted to between 1 and 2%. A series of measurement indicates that the source of carbon black is methane exclusively.

By choosing proper catalysts, plasma–catalytic process can also be used for CH_4 conversion to unsaturated C_2 hydrocarbons (mainly acetylene and ethylene). Since 2005, a series of plasma-catalytic processes based on GD plasma spouted bed has been proposed for CH_4 conversion (Lesinski et al. 1985) by scientists of Faculty of Chemistry, Warsaw University of Technology, Noakowskiego in Poland. In order to improve the selectivity of plasma processes, a hybrid system was proposed, combining plasma activation of the reactants with the use of a solid catalyst that was able to stimulate the reaction course towards the required product. Therefore, most of the excited molecules or radicals may react only in the gas phase, before going into contact with the catalyst. Thus, the distance between the discharge zone and the catalyst's surface may be decisive for the reaction yield. In all of the plasma–catalytic systems selected for examination, C_2 hydrocarbons, besides hydrogen, were the main products of the non–oxidative conversion of methane; however, some amounts of non–volatile substances were also produced, mostly soot. When using alumina–ceramic particles or the catalysts of minor activity (Cu/Al_2O_3, Ni/Al_2O_3, Ag/Al_2O_3), acetylene was the dominating product, with none or only traces of other C2 hydrocarbons, reduction of carbon formation can be achieved. In the presence of Pt/Al_2O_3, however, much more ethylene and ethane were obtained. The share of individual products varied with changes in the discharge conditions, controlled by the reactor input power. In the methane conversion to C_2 hydrocarbons as a function of input power in a homogeneous system and in the presence of tested catalysts, increased conversion of methane to C_2, compared to the homogeneous system was observed for all the catalysts, including the alumina ceramic carrier; the effect of the platinum catalyst was the strongest. At the same time, the soot generation was reduced for all of the catalysts except for Cu/Al_2O_3. The presence of a spouted bed of particles in the discharge volume influences the electrical parameters of discharges. The active catalyst, especially Pt/Al_2O_3, can modify the mechanism of non–oxidative methane conversion and change the final product composition, resulting in increased ethylene and ethane contents and a decreased content of acetylene. The decomposition of CH_4 using a spouted catalytic bed with a gliding arc discharge showed the decline of CH_4 conversion because of the distortion of the discharge by the catalyst. Meanwhile, there was a significant change in the products' selectivity. Acetylene was the main C_2 hydrocarbon product under the gliding arc discharge without a catalyst. Under the Pt and Pd catalyst beds, C_2H_4 and C_2H_6 replaced C_2H_2 through hydrogenation. Pd/Al_2O_3 presented strong resistance to deactivation than did Pt/Al_2O_3, and consecutive hydrogenation from C_2H_2 to C_2H_6 progressed satisfactorily under the plasma condition. The proposed reactor has a potential to improve the selectivity of the plasma process. It is expected that this hybrid system could be used widely where an interaction of plasma process and catalytic reaction is needed, such as decomposition of VOCs, reforming of biogas and reduction of NO_x. The system will also be applicable for plasma–solids reactions including catalysts such as coal gasification.

8.3 Environmental Protection

8.3.1 Abatement of VOCs

The emission of volatile organic compounds (VOC) is an issue of major concern due to severe threat to the environment and human health (Du and Yan 2007). Thus, growing environmental awareness has led to stringent regulations to control VOC emissions and as a result a disposal process with high efficiency, reliability and cost effectiveness is necessary to convert these compounds. Nowadays, catalyst plays a key role in the modern industry for the high efficiency and economic reason. To ensure high efficiency of heterogeneous catalysis, highly dispersed active phase is in need. Currently existed preparation processes of dispersed catalysis are usually a series of combination of different techniques. However, the present status of technologies is far away from preface. Utilization of plasma fluidized bed for the synthesis and pretreatment of the catalyst has become a promising aspect.

Among the processes studied to effectively remove VOC, non–thermal plasma (NTP) coupled with catalyst bed in electrical discharges has many advantages. Over the last decade, many studies have been dedicated to the evaluation of plasma process at atmospheric pressure for the removal of VOC at low concentration (most of the time bellow 1000 ppm) (Fabry et al. 2013).

For over ten years, the study of diphasic processes combining a plasma at atmospheric pressure and a catalyst material has emerged and let to an impressive number of scientific articles in the literature. The original idea of the first studies was to combine the advantages of operating non–thermal plasma at room temperature, their low power consumption, their high chemical reactivity in gas phase (through the production of O, OH, O_3, N_2*, ... species), and an additional catalytic effect, by immersing a catalyst in a plasma technology is the production of many by–products (Du 2014a; Du et al. 2014b), such as benzaldehyde and HCN. To resolve some of these problems, one solution is to couple the plasma with a fixed or fluidized catalytic bed. Such processes operate at room temperature and atmospheric pressure, over a wide range of pollutant types (Shi et al. 2009a, b; Du et al. 2010, 2013; Lu et al. 2014a, b; Ren et al. 2013) and concentrations (Du et al. 2012), with a fast startup [ns time scale (Du et al. 2014c)], and a low power consumption (Kumar et al. 2008). Indeed, the energy injected into the plasma is dissipated through reactive process producing active species (O, OH, O_3, NO_x, N_2 vibrationally excited and metastable states) through reactions involving electrons, rather than heating the gas. These reactive species will initiate removal reactions of the selected pollutant. However, for catalyst with a size of micrometer scale or nanometer scale, fluidized bed is much more proper to exert the synthetic effect since fluidization can enhance the contact of the catalyst and the polluted gas to the largest degree. In the case of acetaldehyde CH_3CHO) polluted air, a process coupling DBD plasma and a fluidized nanostructured silver based bed has shown good performance compared with plasma alone process. The results show that more than 97% of the initial pollutant can be removed at room temperature and with the power injected into the plasma lower than 500 mW and SIE $\leq$ 150 J/L. The CO selectivity was decreased to a large degree (Wu et al. 2010).

Oxidation of toluene in low pressure plasma catalytic fluidized bed reactor pointed out the possibility to study its reactivity without any interference with thermal effects. In this work it was proved that it is possible to destroy aromatic hydrocarbons or to make high added value compounds with the use of catalysts. Under the same experimental conditions type n–semiconductor oxide catalysts (WO_3/NiO) lead to the destruction of toluene while zeolites produce high added value hydrocarbons. The use of mordenite (ZM760) enhances the formation of liquid oxidized hydrocarbons while this of faujasite (YNa) leads to the formation of gaseous and solid products. In the future work the use of WO_3/IViO has to be investigated under thermal plasma conditions but also the effect of rations (copper or cobalt) that enhance the catalytic activity of zeolites (Yan et al. 2013). Previous study investigated that plasma process can effective destruct the toxic wastes containing C and F or Cl, such as CF_4 or $CHCl_4$ effectively and meanwhile avoid the formation of the phosgene and cyanides, which are easily generated during other thermal process. However, second products such HF and CHF are likely to be formed. Plasma process coupled with a $CaCO_3$ or CaO beds provides an elegant solution to destruct the toxic wastes and at the same time, to trap the fluorine atoms lessening the second pollution (Şen et al. 2012; Zhu et al. 1995).

The aim of the present work is the study of the selective oxidation of toluene. Experiments have been performed in a low–pressure (80 Pa) fluidized–bed of catalyst plasma reactor. Depending on the nature of the catalyst, the oxidation of toluene leads to two kinds of products. The total destruction of aromatic molecule with formation of carbon oxides is obtained with a type N semi–conductor catalyst while high added value molecules (phenol, cresols, etc.) are formed when zeolites (mordenite or faujasite) are used in the fluidized–bed. Depending on the catalyst used, such a process can find applications to destruction of toxic molecules or the formation of high added value compounds under "limited temperature" conditions.

Except for plasma–catalytic fluidized bed, thermal plasma fluidized bed may be a better choice for the removal of the pollution. For example, Elizabeth et al. proposed an interaction of microwave energy with fluidized carbon granules for in the remediation of airborne organic pollutants. Larger plasma density generated by microwave discharge and temperature around 250 degrees centigrade through the dielectric heating the activated carbon with microwave work together and offer ethyl acetate. Due to their excellent adsorption and specific catalytic properties, carbon materials have been widely used as adsorbents and can give rise to the possibility of catalytic oxidation reactions in the oxidizing environment (Flamant 1990; Matsukata et al. 1992). It is widely known that electrical discharge plasma produced oxidizing environment containing H_2O_2, O_3 and other species in the reactor, which is desirable for heterogeneous catalytic reaction of activated carbon (AC). Grovender et al. (2001) and Mochizuki et al. (1993) initially introduced activated carbon into pulsed streamer–like plasma degradation system for phenol oxidation leading to enhancement of the overall removal of organic as compared with the electrical discharge alone. In these combined processes, activated carbon samples can physically and chemically adsorb both organic compounds and oxidizing molecules (O_3 and H_2O_2) due to its high surface area and abundant surface functional groups (Yang et al. 2006).

8.3.2 Control of NO_x

Nitrogen oxide (NO_x) control remains one of the most important technical challenges till today. The decomposition of NO_x by the nonthermal plasma process has been investigated intensively. In the non–thermal plasma process without additives, NO can be effectively converted to NO, but NO cannot effectively be reduced to N_2.

Part of the NO will be converted into NO, HNO (or NO), and NO as indicated in reactions. For these reasons, NO reduction using the nonthermal plasma technology alone has limitations. A three–way catalyst used for automobile emission control, such as the copper–coated zeolite catalyst (Cu–ZSM5), has been investigated combined with nonthermal plasma, but NO conversion was insignificant, even at the elevated temperature of 250 °C. When the pulsed corona discharge plasma was combined with the addition of hydrocarbon and the catalysts, conversion of as much as 90% of NO was reported (Shi et al. 2009a, b; Du et al. 2010), but NO generation was not discussed. Also, the use of water vapor and a Ca(OH) catalyst showed some promising NO conversion in the range of 70% (Shi et al. 2009a, b).

In recent studies, plasma fluidized bed has been taken to modify Co–EFR catalyst, Pd/γ–alumina and Ag/γ–alumina catalysts (Al–Shamery et al. 2009) for the $DeNO_x$ processes. It was observed that the plasma treated catalyst showed a excellent $DeNO_x$ activity compared with those treated by conventional calcination (500 °C, 2 h in air), especially in the lower temperature range. Uniform and chemically active atmosphere in the plasma fluidized bed lead to the formation of different reactive species and active sites accounting for the interesting catalytic feature in the low temperature range and work for a satisfactory $DeNO_x$ performance, several suggestions for further investigation are given: the radio frequency generator will be replaced by a microwave source, which presents the advantage to create plasma with higher density of energy leading to lower treatment times and better fluidization.

8.3.3   Sterilization of Food

Recently, plasma fluidized bed has also been found with great potential for the inactivation of microorganisms on granular materials by Denis Butscher in Switzerland (Butscher et al. 2015), which is meaningful for the food safety due to more restrictive food laws and higher quality. A low pressure circulating fluidized bed reactor was constructed for the treatment of granular materials. In the riser tube of this reactor inductively coupled RF–powered argon/oxygen plasma was ignited, which was used to inactivate microorganisms on the particle surface. With this setup, the amount of B. amyloliquefaciens endospores, arti ficially deposited on wheat grains, could be reduced by more than two logarithmic units within 30 s of effective treatment time. Higher plasma power input intensi fied the discharge and increased the axial extend of the treatment zone, and spore inactivation seems to be more efficient at higher oxygen to argon admixtures (10% compared to 5%). Thermal inactivation and wheat grain degradation could be excluded, based on energy influx measurements and the solution of the heat equation. In addition, Farinograph, Extensograph and Amylograph measurements showed no negative effects of plasma treatment on flour and baking properties expectations.

8.3.4 Plasma Modified Catalyst for Water Purification

Similar to the removal of NO_x, although plasma fluidized bed have not been used directly for polluted water purification, catalyst such as TiO_2 and other photo–catalyst generated with various plasma fluidized bed have been successful for wastewater treatment. To illustrate, pollutant such as phenol or methylene blue can be efficiently removed. Catalyst modified in plasma fluidized bed is characterized of low cost and more controllable features. What's more important, the wettability and the stability of the used catalyst in water can easily be modified in the plasma fluidized bed by various operations such as oxygen fictionalization so that the activity and lifetime of the catalyst in water can be increased for a better performance. However, the fundamental interaction of microwave energy with fluidized carbon granules and a potential application in the remediation of airborne organic pollutants has not been studied extensively.

8.3.5 Solid Waste Treatment

Waste management is an important issue in both developed and developing countries nowadays (Du and Yan 2007; Du et al. 2008, 2014d; Shi et al. 2009a, b). Various plasma fluidized bed offers a good way to helps deal with waste management problem. Plasma fluidized bed for solid waste treatment can be generally divided into three ways: gasification/prolysis, vitrification of hazardous materials, and the chemical degradation of the hazardous component for soil. As presented in the last section, gasification/prolysis of solid waste could not only help solve the solid waste pollution, but also transfer chemical energy in solid waste to useful heat. Presently, the strong expansion in the world of numerous plasma gasification plants (projects and operational plants) shows clearly that a step has been taken and in the future, plasma gasification will play a significant role in the field of renewable energy. Concerning the advantages of the waste gasification by thermal plasma, the role of the plasma treatment is twofold: it allows, on the one hand, a significant purification of gas by limiting the production of tars and on the other hand, producing a synthesis gas enriched in hydrogen (water–gas shift reaction). Plasma methods have also the advantages to be able to operate at high temperature and to be retrofitted to existing installation. Such a temperature in plasma can allow synthesizing or degrading chemical species in some conditions unreachable by conventional combustion and can greatly accelerate the chemical reactions. Thermochemistry of combustion does not allow precise control of the enthalpy injected into the reactor. Plasma process allows an easiest enthalpy control by adjusting the electrical power. The reactive species produced by the plasma, such as atomic oxygen and hydrogen or hydroxyl radicals, is an additional advantage for the use of plasma and enhance strongly the degradation of the tars with greater efficiency than conventional processes. This is a growing market and the efficiency of the waste gasification by plasma seems to be validated but the economic viability of this technology must be proven before to be accepted by the industry.

Plasma vitrification is a technologically advanced and environmentally friendly method of disposing of waste, converting it to commercially usable by–products. Concerning the development and the operation of the plasma technologies on the energy market, presently, the technical feasibility and economical viability of plasma vitrification technologies have been demonstrated for a large range of hazardous wastes but it is not totally the case of plasma gasification technologies for the disposal of MSW at an industrial scale (Bao et al. 2006). This process is a drastic non–incineration thermal process, which uses extremely high temperatures in an oxygen–starved environment to completely decompose input waste material into very simple molecules. The intense and versatile heat generation capabilities of plasma technology enable a plasma gasification/vitrification facility to treat a large number of waste streams in a safe and reliable manner. The by–products of the process are a combustible gas and an inert slag. Plasma gasification consistently exhibits much lower environmental levels for both air emissions and slag leachate toxicity than other thermal technologies.

Expect for slag treatment, vitrification can also be used for many other hazardous materials disposal, such as waste hospital materials, PCBs, waste electrical materials and so on.

In conclusion, a large number of good points can be achieved by the implementation of the gasification/vitrification process for waste treatment after the optimization of the operating parameters for every different waste stream (Ren et al. 2013; Li et al. 2010). Actually, the vitrification of the solid waste is always with the role of gasification.

The third way for solid waste treatment was the chemical degradation of hazardous component to detoxication, such as soil remediation and the abatement of fly ash.

In previous study, Du et al. found that Vortex gliding arc plasma (VGAP) can effectively degrade PCDD/Fs in fly ash, whether in oxygen, air, or nitrogen. The degradation characteristics of various PCDD/Fs congeners vary with the atmosphere used. Total degradation efficiencies range from 54.9 to 66.8% on a mass basis and from 60.7 to 73.3% on a toxicity basis. After thermal desorption a small amount of PCDD/Fs may transfer from fly ash to the exhaust gas and dechlorination can also be observed. Also the fly ash surface undergoes obvious changes: needle–shaped crystals, pores, fragments, and especially high–temperature melting all are observed after the treatment.

On the basis of experimental results some degradation pathways and a mechanism of decomposition of PCDD/Fs in fly ash are proposed. PCDD/Fs can be degraded by active groups produced by VGAP, including high–energy electrons, excited atoms (N, O, etc.) and molecules (O_3 and NOx), as well as free radicals (OH and NOx) UV illumination. In addition, dechlorination mainly takes place in nitrogen, but under conditions of oxidative degradation occupies an important role. Degradation end products are obtained including CO_2, CO, CH_4 and HCl, etc.

Also, in 2014, Du (2014f) in China also invented a special GD plasma fluidized bed for soil remediation and the inventor found that this special reactor has a significant role in solid waste treatment including MSW, soil and other kinds (Sabat et al. 2015; Mendes et al. 2008; Chen et al. 2009; Yan et al. 2012; Li et al. 2014; Wang et al. 2006; Bai et al. 2005; Bieri et al. 2007; Lu et al. 2014a, b). A novel non–thermal plasma fluidized bed (PFB) was developed for the remediation of Phenanthrene (PHE) contaminated soil, as can be seen in Fig. 8.3. After 25 min treatment, 95% of the PHE was removed with an energy density 5960 J/g soil, air flow rate 20 L/min and soil moisture 10%. The effects of the solid bed location, energy density, the flow rate and the carrier gas were explored. It was found that decreasing the distance between the solid bed and electrodes properly and

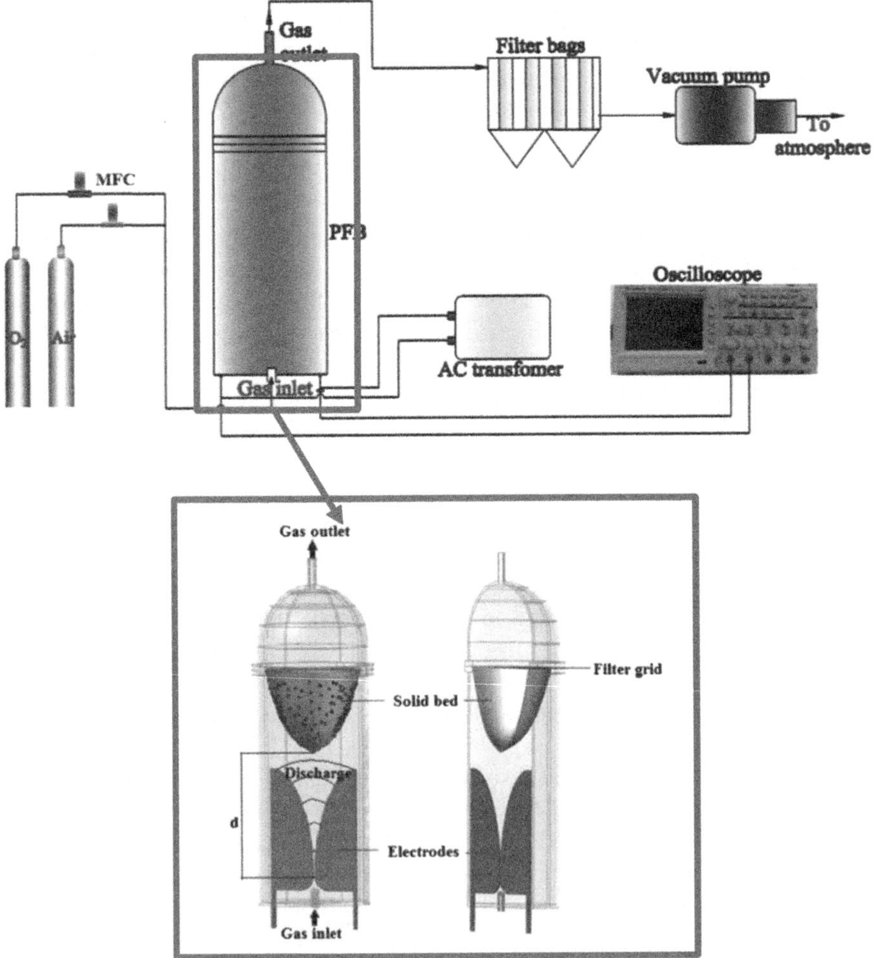

Fig. 8.3 GD plasma fluidized bed for soil remediation

increasing the input energy density were favorable for PHE degradation. There existed appropriate flow rate and soil moisture for a higher efficiency. By means of various methods, the role of the active species on the degradation efficiency and the products formed during discharge was analyzed. Possible mechanism of the PHE degradation by the PFB was proposed. Furthermore, a brief comparison with other competitive processes was performed, indicating that the proposed PFB was a promising alternative process for PHE contaminated soil remediation.

Therefore, plasma fluidized beds have been proved to have a huge market high efficiency, reliability and cost effectiveness for soil, water, and gas pollution. And further modification and improvement is for the scale up for environmental application.

8.4 Material Field

8.4.1 Surface Activation and Functionalisation

Functionalizing nanoparticle surfaces is essential for achieving homogeneous dispersion of monodisperse particles in polymer nanocomposites for successful utilization in engineering applications. Functionalization reduces the surface energy of the nanoparticles, thereby limiting he tendency to agglomerate. Moreover, reactive groups on the surface can also participate in the polymerization, creating covalent bonds between the inorganic and organic phases.

The wettability of a substance is also directly linked with the intrinsic dissolution rate (Du et al. 2010). Thus, the dissolution behavior of a powder can be tuned in a certain range by only creating some functional groups on the particle surface.

Apart from that, the tendency to agglomerated, contact angle (wettability), flow-ability, resistance is always the research focus of various investigations. Previous study suggested that by using MW plasma fluidized bed with nitrogen as plasma gas and carrier gas, it was possible to the increase the wettability of a PE powder.

MW plasma with 2.45 MHz is the common standard in research ad industry to avoid interference with telecommunication applications. The generation of the MW plasma depends strongly on the pressure and gas composition. Among the application of MW plasma fluidized bed, remote nitrogen plasma reactor coupled with fluidized bed has been extensively investigated for the treatment of the polyethylene (Lesinski et al. 1985) and the typical description of the remote nitrogen plasma fluidized bed showed that by using MW plasma fluidized bed with nitrogen as plasma gas and carrier gas, it was possible to the increase the wettability of a PE powder. The contact angles with water, higher than 90 for the original PE powders, were equal to 0 after a treatment for 25 min with 0.75% of oxygen added to the nitrogen gas. Also, the flow-ability of the powder was found to be slightly influenced due to agglomeration resulted from the increase of the wettability. By XPS analysis, Mutel et al. (2003) proved that the oxygenated and nitrogenated functions are incorporated after plasma treatment. The mechanism of the change of the wettability was ascribed to the fixation of the new chemical function. Authors found that radicals generated in the discharge can initiate the fixation of the new chemical functions on the surface but cannot directly affect the wettability of the surface. To illustrate, imine function can be partly hydrolysed to form amide group. Powder flowability (Allah and Whitehead 2015) was another important aspect of various industrials applications. The presented process is much faster compared to conventional techniques aiming at increasing flowability for instance by admixing nanoparticles.

Besides oxygenated and nitrogenated functionalization, fluorinated and chlorinated functionalisation can also be achieved inside plasma fluidized bed using CF_4 and CCl_4 as operating gas. Plasma surface fluorination upon the particles can be achieved through the formation of functional groups such as $CHF–CH_2$, $CHF–CHF$ and CF_3 groups. And that by means of the change of duration periods of the plasma modification, the plasma power input and, the operation composition and the velocity and some other operating parameters, the composition as well as the intensity of various functionalization can be determined.

Nevertheless, in the MW range, electrons are accelerated and collisions are induced less efficiently as the RF range due to the higher frequency, as a result, RF plasma fluidized bed occupied an important role in surface modification, which can be operated under both low–pressure condition and atmospheric pressure condition.

The hydrophobic surface can be transformed to hydrophilic of HDPE powder and the oxygen functionality is formed including $C=O$ and $C(O)O–$, which reach 12 and 8% respectively of the total carbon elements by the oxygen plasma treatment. The CFB reactor outperforms 3.4 times to obtain the similar level of hydrophilicity compared to that in a bubbling fluidized bed based on the composite parameter (Emome and Jurewize 1999; Tsukada et al. 1995). The effects of the RF plasma power, the gas velocity, the duration time on the surface modification have been investigated. And generally, the oxygen functionality decreased with the increase of the oxygen flow rate, but increased with the RF plasma power.

The oxygen molecule can be dissociated, ionized or activated in the discharge. By this mean, polar groups such as hydroxyl or carboxyl are formed on the surface of substrate particle. These functional groups increase the surface free energy of the powder and thus, enable low contact angles even with polar liquids such as water. The plasma–assisted wettability improvement of powders has extensively been studied for polymers e.g. in (Shi et al. 2009b). By this way, polar component such as OH, COOH etc. can be generated upon the powder surface and therefore, the polarity and other character can be changed significantly. The residence time in the PDR is about 0.1 s, while adequate mixing can take several minutes or even hours in the case of cohesive and fine–grained powders.

In addition, no segregation can occur if the nanostructures are directly bond to the substrate material. The implementation of the subsequent surface activation increases the number of potential applications in industry since also products with poor wetting or dissolution characteristics (vitamins, pigments, pharmaceuticals, nutraceuticals, or food additives) could benefit from the process in this form.

Generally, the majority of the RF plasma fluidized bed reactor and MW plasma fluidized bed were operated under the low pressure condition. Low–pressure plasma have a dominant and long–established role because of their unmatched capability of providing a vast array of chemically active species, low gas temperature and uniform reaction rate over a relatively large area. Yet, a major limitation of this kind of plasma is the requirement for relatively expensive and complicated vacuum operation. In recent years, much attention has been paid to nonthermal atmospheric pressure plasma and some atmospheric pressure discharge sources have been developed.

8.4.2 Pecvd

8.4.2.1 SiO$_x$ Thin Film

Among various technologies, the deposition of a thin film on the particle surface seems to be an elegant option to attain desired surface features, which is expected to lead to innovative, flexible and cost–effective particle coating processes for uniform coating of powders. The plasma deposition process can be divided into two groups: plasma enhanced chemical vapor deposition and plasma polymerization. The principles of these two processes are similar except that the PECVD is used for processes where the deposited films are of rather inorganic character, whereas the deposition of organic films is usually called plasma polymerization.

PECVD is a rather hot field for research on the application of plasma fluidized bed. Up to now, various plasma fluidized beds have been used for this application, ranging from DC torch plasma fluidized bed, RF, microwave, DBD to non–thermal arc plasma fluidized bed. The most typical deposited materials can be classified as silicon layers, metal layers such as TiO_2 as well as the carbon layers.

Silica is an attractive deposit material due to its excellent thermal and chemical stability, and it's deposit principle is shown in Fig. 8.4. According to literature, potential applications in the field of PECVD of SiO$_x$ on particles are corrosion protection of pigments (El–Naas 1996; Matsukata et al. 1992; Spillmann et al. 2007; Yamamoto 1997), diffusion barriers of pharmaceutical powders to retard the active agents (Sarjeant and ROY 1967; Bretagnol et al. 2004), enhancement of chemical resistance of powders (Gomez et al. 2008) and wettability modification of bulk solids (Du et al. 2012). To achieve a three dimensional and uniform coating, a plasma fluidized bed is in need to offer desired heat and mass transfer between the reactive species generated in the discharge flow and the substrate particles by means of intensive mixing. Furthermore, it is important to govern the mean residence time of the particles in the plasma, e.g. to control the thickness of a deposited layer on the particle surface.

Fig. 8.4 Schematic diagram of the deposition mechanism of the APTPECVD process

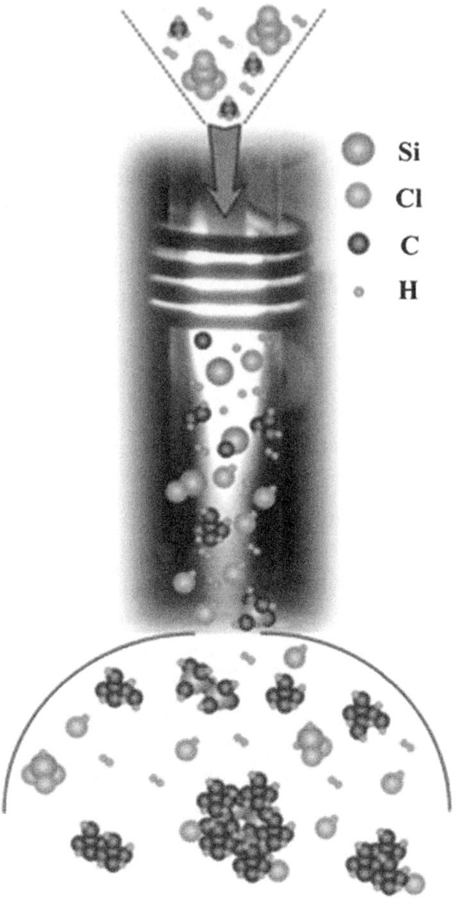

Due to its special advantages, numerals investigations have been carried out for the deposition of silicon layer on the particles. Fluidized bed reactors are suitable to coat particles with sizes between 50 and 500 μm. For stable discharge and full utilization of the generated species, it is often necessary to operate a fluidized bed reactor at vacuum conditions below 10 mbar. Up to now, there are also scientists making efforts to introduce an atmospheric plasma fluidized bed for the PECVD, especially DBD plasma fluidized bed (Li et al. 2010). However, due to the fact that the plasma density of DBD process is much lower than that of the RF or microwave plasma, the deposition rate the utilization efficiency of the monomer gas in beyond satisfactory.

Coatings of TiN, SiN and SiO were deposited at low temperature and atmospheric pressure on fluidized particles of silica and corundum is a plasma jet fluidized bed reactor. The immersed gliding arc permits to create inside the bed volume the excited species, which react to form the deposit layer on the particle surface. To a certain extent, the coating composition may be adjusted from oxides to nitrides, although pure nitrides are more difficult to obtain because of residual oxygen pollution inside the reactor. Layer thicknesses ranging from 2 to 10 mm were obtained, which correspond to growth rates in the range 5–25 mm/h (Karches et al. 1999).

According to previous studies, Silicon–containing organic compounds are preferentially used as monomers in low–pressure plasma deposition of silica (SiO_2) and silica–like (SiO_x) thin films. Since non–stoichiometric SiO_x films have varying hydrocarbon content, the proper empirical formula would be $SiO_xC_yH_z$. The advantages of organic silicon compounds are their availability, liquid state, volatility at room temperature, safe handling and low costs (Shi et al. 2009a). Besides silane (SiH_4), tetraethoxysilane (TEOS) and others, hexamethyldisiloxane (HMDSO) is a commonly used monomer for the PECVD of SiO_2 or SiO_x films (Matsukata et al. 1992). Due to the high relative velocity, there is an intense gas–solid contact resulting in an enhanced mass and heat transfer. Therefore, fluidized bed reactors provide an enhanced heat transfer to the reactor wall (good temperature control), a uniform temperature distribution in the bed, and a smooth particle agitation (reduced material attrition). To illustrate, Salt crystals (NaCl) with a mean diameter of 551 µm were treated in a low–temperature plasma fluidized bed reactor by Bayer et al. (Karches et al. 1999). The power was coupled by a MW plasma source by varying the power between 690 and 1590 W and the process gas consisted of HMDSO, O_2 and Ar. It was shown that PECVD processes in fluidized bed reactors can be operated at larger scales up to bed masses of 3 kg. Due to the SiO_x coating on the particle surface, the salt crystals showed hydrophobic properties after the plasma treatment. However, deposition processes in fluidized bed reactors are faced with the problem of heat dissipation as well as the achievement of a uniform coating (El–Naas 1996). Thus it is important to have a narrow size distribution of the particles, in order to avoid particle segregation in the bed, which would result in non–uniform coating. Also, heat dissipation is another problem.

The proposed spouted–bed gliding arc discharge method produced PMMA particles with fine zinc coating layers just after 5 min of coating operation. The amount of zinc coating increased with time but reached equilibrium after 20 min when the coating and peeling off rates would become equal. As coating duration increased, the coating thickness increased but the percentage surface coverage remained relatively constant. On the other hand, when the zinc/PMMA mass ratio increased, both coating thickness and surface coverage increased. The non–uniformity of surface coverage may be attributed to frequent inter–particle collisions and weak adhesive strength between the coating layer and PMMA surface. As a result, electrical conductivity was imparted to the nonconductive PMMA. The novel approach of integration between the gliding arc discharge and the spouted bed condition has high potential for rapid particle coating.

In order to obtain lower temperature and pressure gradients and a narrower residence time distribution, Karches et al. (1999) introduced the plasma circulating fluidized bed reactor for PECVD on particulate materials. In the study, NaCl crystals (210 µm) were coated with thin silicon oxide film and the deposition rate of

1 µm/min was measured. In circulating fluidized bed, the intensive heat transfer of particle–particle and particle–wall can efficiently avoid the horizontal temperature gradients and hot spots. It was shown that the temperature of the particles can be controlled by the solid mass flow to less than 100 °C. By means of increasing the circulation number, the high homogeneity of the process and then of the coating can be achieved. Borer and Rudolf von Rohr (Du et al. 2010) investigated the structure of the growing SiO_x film deposited in a $HMDSO/O_2/Ar$ MW plasma on NaCl (280 µm) and silica gel (300 µm) particles by using the reactor set–up of Karches et al. (Karches et al. 1999). On the smooth NaCl crystal surfaces, a dense coherent film was achieved. However, the film constitution was disturbed by nodular structures which are caused by dust particles. These particles are embedded in the coating during the deposition process and due to the shadowing effect, the defects are propagating in the film. By contrast, no dense and coherent films could be deposited on the rough surface of the silica gel particle because of the initiation of columnar growth on the primary particles of the substrate material. In a subsequent work, Borer et al. (Rohr and Borer 2007) investigated the effect of substrate particle (silica gel, 300 µm) temperature on the growth structure of the deposited film. For this, the particle temperature was varied between 61 and 195 °C by changing the MW power input. It was shown that at lower temperatures, columnar structures dominate the film morphology. By increasing the particle temperature, the columnar growth effects were reduced while an increase of the coherent film fraction was observed. According to the structure zone model proposed by Movchan and Demchishin, this finding was established due to the higher adatom mobility at elevated surface temperatures. Jung et al. (Jung et al. 2004) used an atmospheric DBD in a circulating fluidized bed (CFB) reactor. For this purpose, ring–shaped electrodes were arranged around the quartz glass riser which served simultaneously as the dielectric barrier and reactor confinement. The alumina particles (60 µm) were treated at different power inputs at a frequency of 13.56 MHz. A gas mixture of TEOS, O_2, Ar and He was fed to the reactor to coat the substrate particles with SiO_x. According to SEM analysis, the thickness of the film was in the range of 2–4 µm. In the identical experimental set–up, it coated alumina powder (100 µm). In contrast to the previous work (Matsukata et al. 1992), both monomers TEOS and HMDSO were investigated as a precursor. The water wettability on the plasma–treated alumina could be adjusted from hydrophobic to hydrophilic by varying the applied power. Furthermore, the effect of discharge power and O_2/monomer ratio on the organic content of the film was analyzed.

The SiC nanocrystal is a promising material with potential applications in electronics, photovoltaics and biological fields. In this work, SiC nanocrystals were fabricated using an APT–PECVD process with $SiCl_4$ as the Si source and CH_4 as the C source. The thermal–plasma–enhanced CVD process and product properties were thoroughly studied. The SiC nanocrystals produced were covered by carbon 1 ms and embedded in the network formed by graphite and a–Si. The deposition rate of the thin film increased with increasing $SiCl_4$ input rate and reached a maximum value 70 nm/s. The SiC nanocrystal has an average grain diameter in the range of 21–33 nm. The effect of $SiCl_4$ input rate on the surface chemical compositions of the samples was studied, and the C atomic fraction and SiC mole fraction of the samples were calculated. An OES diagnostic was carried out to better understand the effect of $SiCl_4$ input rate on the deposition process. The concentrations of atomic Si and C in the gas phase were found to be the main factors affecting deposition rate and product properties. Based on these experimental results, a simple deposition mechanism in the APT–PECVD process was deduced, indicating that SiC nanocrystals were formed by the assembly of atomic species formed by thermal plasma decomposition of Si and C precursors.

The application of a circulating fluidized bed for PECVD on powders is restricted to sufficient high particle size, i.e. mass, in order to exert the necessary momentum to the powder bulk, especially in its separation from the conveying gas. Furthermore, for very small, cohesive particles which tend to stick together, the fluidization of the powder is hard to achieve and the solid tends to stick on the reactor walls. Moreover, if fast treatment and short–reaction time is needed, the circulating fluidized bed is improper due to its broad residence time distribution for a small number of circulations. To overcome these problems, Arpagaus et al. (2005), Bashlai et al. (1972), and Aranovich et al. (1973) used a downer reactor for the surface activation of polymer powders. The concept of this reactor will be explained in more details in Chap. 2. Compared to the up–flow in the riser, the radial velocity profile in the downer is much more homogeneous and approaches plug flow. Further, they have shown that the axial solid mixing in the flow direction against the gravity (riser) is much larger than in the flow direction of the gravity (downer). These findings were explained by the domination of different mixing mechanisms in the downer and riser. The solid mixing in the direction of the gravity is dominated by the dispersion of dispersed particles, while the dispersion of particle clusters is prevailing in the up–flow. Since the dispersion due to particle clusters is much more significant, axial mixing is more pronounced in the riser. These results were confirmed by Zhang et al. who investigated the velocity profiles and distributions of the solid holdup in a riser–downer system. These indices are defined as normalized standard deviations of the particle velocity and the solid holdup in the radial direction of the tube. It can be seen that the radial flow structure in the downer is much more uniform than that in the riser. Furthermore, the downer shows more uniform velocity profiles in axial direction. Based on these observations, it was concluded that the homogeneous radial flow structure in the downer favors a narrow residence time distribution (Krawczyk et al. 2009).

8.4.2.2 Diamond–like Coating

Amorphous hydrogenated carbon layers also called diamond–like carbon coating. Diamond is one of the most attractive materials because of its excellent properties, such as high hardness, good thermal conductivity and chemical inertness. Although flat diamond films have many applications, others require particles which are uniformly covered with diamond coatings, for example, uniform diamond powders and diamond coated fibers, which was hard for conventional CVD process. DC plasma fluidized bed and microwave plasma fluidized bed have been proved to be good alternates for the deposition of diamond films from the gas precursor on various substrates, such as Si and SiO_2 powder (Dasan et al. 2016; Uemura et al. 2003; Du et al. 2013). Generally, CH_4 and H_2 can be used as processing gas, and inside the fluidized bed, transport of reactive radicals (such as H) to the substrate particles very efficient and uniform. Much different from conventional coating techniques, for which deposition of diamond–like coating on substrate usually requires very high temperature, in plasma fluidized bed, the diamond coating was found to occur in the plasma tail with a very low temperature, inferring that the formation of diamond deposition can be ascribed to the presence of reactive species (Dasan et al. 2016). Also, the introduction of oxygen into the process gas can significantly enhance the deposition of diamond, for it can increase the reactive radicals and lower the need for high temperature. Therefore, for higher quality and quantity of synthesized diamond coating, a higher plasma density needed, microwave plasma fluidized bed and NTP plasma technology such as GD plasma fluidized bed may provide enough reactive particles for reaction. Also, the enhanced contact of the plasma gas and the substrate particles may be another way to increase the conversion efficiency.

In a plasma environment, the synthesis of nanodiamond crystals and ultrananocrystalline diamond films can be implemented under relatively low (well under atmospheric) pressures and moderate process temperatures. However, to mitigate the higher energy cost of the synthesis of the more non–equilibrium nanodiamond structures, plasma of relatively high energy densities and species concentrations (e.g. microwave plasmas) are commonly used. The plasma–assisted synthesis of nano–diamonds is a relatively well established and advanced process developed over the last couple of decades (185,186)(Liu et al. 1996; Pajkic and Willert-Porada 2009). Diamond nanocones and nanorods are other examples of plasma–produced sp^3 carbon nanostructures. (Kim et al. 2009; Sanchez et al. 2001)

8.4.2.3 Metal Content Coating

Pajkic in University of Bayreuth in Germany reported an innovative process for deposition of thin metallic and metal–nitride coatings (Pajkic and Willert-Porada 2008; Du and Xiao 2014), using a fine solid metal powder is used as "target powder", serving as a source of metal vapor at atmospheric pressure and temperature below the melting temperature of the metal. We coin for this new process the term "fluidized bed arc–PVD coating". During the residence time in the sustained microwave plasma, metal atoms evaporate from the target powder due to the charged inert gas particle bombardment in the plasma and condense at the substrate particles surface. We coin for this new process the term "fluidized bed arc–PVD coating", because of major difference to the well known arc–PVD–process, which utilizes fixed, stationary, and compact targets (Sakano et al. 2001; Sarjeant and ROY 1967). The novelty of the process described here is that both the substrate and the target are particulate materials, fluidized together at the same time, so that the system behaves as a PVD with multiple targets and substrates. Moreover, the residence of this fluidized bed in microwave plasma enables activation and low temperature processing known only from arc–PVD, which has even higher restrictions with respect to the target–substrate arrangement as compared to PVD. This suggests that the activation by charged particle bombardment from the plasma, as in case of arc–PVD, is crucial for this process. It enables simplification and cost reduction of the CVD fluidized bed coating process, because no dangerous, expensive and difficult–to–handle metal precursors have to be used, and the off gas treatment is also much easier. Further cost reductions include also elimination of the complicated and expensive pumping and trapping subsystems, since no vacuum is needed for this process.

One of many applications of functional coatings is catalytic films on law cost substrate particles. TiO_2 coated glass beads are used for instance as photo catalyst for decontamination in wastewater treatment (Du et al. 2014d). Circulating fluidized beds coupled with RF or MW plasma have been successfully utilized for the deposition of catalytic TiO_2 (Jung et al. 2004) film upon different bulk particles (NaCl crystals, glass beads and silica gel etc.).

The anatase–TiO_2 thin films were prepared by using titanium–tetraisopropoxide (TTIP, $Ti(O–i–C_3H_7)_4$) and oxygen or Tetrakis (tert–butanolato) titanium (Ti $(OtBu)_4$ and oxygen. For a larger contact area for particles and plasma zone, argon and helium can be mixed as process gas for a larger stable plasma discharge zone inside the riser tube of the reactor. PECVD in the circulating bed for TiO_2 deposition seems to be very energy saving. It can be claimed that crystallization into anatase phase on powders can be achieved without any post–treatment in a CFB reactor. With the plasma input power = 350 W, deposition temperature = 250 °C, mass velocity of TTIP = 0.4 g/min, argon concentration = 8.18 vol%, oxygen concentration = 3.6 vol% and the treatment time = 2 h, TiO_2 film is dense, uniform, with less impurity can be successfully deposited on the silica gal particles. By this way, TiO_2 content of 85% was reached (Flamant 2009). It is also found that the products can be directly used for photocatalytic process. With the obtained TiO_2–deposited powders and silica gel powders, the photocatalytic activity to decompose methylene blue aqueous solution, excellent catalytic ability can be achieved.

What's more important, the thickness of the TiO_2 film can be easily controlled by the variation of the operation time. TiO_2 film deposited in circulating fluidized plasma bed.

Gook Hee Kim in Korea has prepared the TiO_2 thin films on powders in a CFB reactor by the PECVD method and the deposited films PECVD in the circulating bed for TiO_2 deposition seems to be very energy saving and is able to offer dense uniform TiO_2 film with a thickness controllable by the change of the operation time (Du 2014). Morstein et al. (2000) and Karches et al. (1999) have prepared the TiO_2 thin films on powders in a CFB reactor by the PECVD method and the deposited films were calcined at 723 K to maintain the catalytic activity (Du et al. 2014a, b). Whereas, in the present study, it can be claimed that crystallization into anatase phase on powders can be achieved without any post–treatment in a CFB reactor.

A novel gliding arc discharge method using zinc particle as a solid precursor was used to coat poly methyl acrylate (PMMA) particles to provide an adjustable electrical conductivity. By using a combination of spouted bed condition and gliding arc discharge, a rapid coating performance was achieved. After 5 min of coating process, PMMA particles were observed to be covered by a fine zinc layer. Similarly, metallic coating can provide other preferable properties, such as electrical conductivity, thereby enabling a non–conductive polymer to gain significant electrical conductivity. The electro–conductive polymer can be applied in many other fields, including battery electrodes, stretchable electronic and medical devices.

In fluidized bed processes utilization of microwave plasma for coating particles has been of major interest since almost 20 years, up to now without break–through towards commercialization, mainly due to problems with reactor scale up.

MW (e.g. carbon fibers, porous carbons), SiC, WC, TiC and finely divided metals. In this case selective microwave heating of the particles could reduce coating of the reactor wall and homogeneous nucleation, improving the yield and purity of the coated fibers.

A wide variety of plasma polymers having various chemical functionalities have been obtained through RF plasma polymerization of numerous monomers. Carboxylated surfaces have been produced by plasma polymerization using monomers such as acrylic acid (Li et al. 2010). However, the carboxyl functionality is easily cleaved off, presumably as CO_2, which reduces the deposition rate. The ability to withstand dissolution is important to prevent false negative results from the underlying substrate during cell adhesion or biomolecules immobilization studies (Li et al. 2010; Attri et al. 2013). The addition of 1,7–octadiene to acrylic acid or the use of higher powers in plasma polymerization of acrylic acid can leads to an increase of the stability of the coating (Attri et al. 2013; Heberlein and Murphy 2008; Pfender 1999; Nezu et al. 2003). To produce a surface containing high density of COOH functions for the immobilization of biomolecules at the surface of PS beads deposition of a thin layer of polymerized acrylic acid on polystyrene beads. The effects of plasma pretreatment, plasma power and plasma deposition time were studied to control the surface density of COOH groups, also to increase the stability of the coatings upon water washing.

Also, the peak corresponding to p–pÃ shake–up satellite reappeared. These results demonstrated a significant elimination of PPAA coatings after washing with water. Indeed, the increase of the input power rendered the deposit less soluble due to cross–linking reactions (Nezu et al. 2003). This result is quite surprising, as an increase of power should lead to higher level of AA monomer fragmentation, thus enhancing the ICO (519.8 nm)/IAr (750.4 nm) ratio. However, we have clearly observed that the use of higher powers lead to an increase of the plasma volume, which could consequently enhance the residence time of PS beads in the plasma region. It was possible to obtain a stable PPAA coating resistant to washing by combining the use of an argon pretreatment of PS beads and optimized plasma deposition process parameters. Our results have clearly evidenced the need to control the density of the reactive nucleation sites created on PS beads, as they could play an important role on the adhesive properties of the further deposited coating, as well as the coverage of the PS beads by PPAA coating. There is also a strong need to control the level of the monomer fragmentation since the formation of low molecular fragments on the surface due to a higher power or residence time could favor the formation of an unstable coating. Chen a nanolayer biofilm of polyacrylic acid was uniformly coated on the surface of magnetic nickel nanoparticles (NPs). The thickness of the biofilm was about 2 nm and the discharging conditions affected the density of the carboxyl group obviously. The PAA acting as an adhesion layer was used to immobilize the antimicrobial peptide LL–37, to kill the bacteria of *Escherichia coli* (*E. coli*), and the results indicated that the modified nickel NPs immobilizing a certain concentration of LL–37 could kill the bacteria effectively. Chen's teams (Ma et al. 2013) used magnetic chitosan Fe_3O_4NPs as an in vivo drug delivery system for magnetic resonance imaging monitored targeting therapy, and the method can effectively attenuate carcinoma cells both in vitro and in vivo.

Most of these materials are designed to perform one task (passive nanostructures), but in the near future, nanostructured materials with multifunctional capabilities (active nanostructures) will be produced at high rates (Du and Yan 2007). Among all candidate materials for use in the upcoming wave of nanotechnology, nano–particulate metallic materials stand out because of their many potential applications, such as an environmental friendly and alternative method, plasma fluidized bed technology could be applied to synthesize nanoparticles. Various plasma fluidized bed technologies can be applied in nanoparticle manufacture instead of flame technology, because of their ability to decrease CO_2 and pollutants release. Additionally, plasma technology is highly flexible in selecting precursors, plasma gas and reaction temperature. For example, titanium nanoparticles can be formed from plasma in a mixture of titanium tetrachloride, argon and hydrogen. Finally aerosol particles are often charged and repel each other in plasma, and it is of bene fit to the formation of ultra-fine particles. They often involve high energy consumption and need complicated cooling systems, resulting in low energy efficiency. Another disadvantage of thermal plasma devices is that their electrodes are easily eroded at high temperatures when oxygen exists. Additionally, it is difficult to treat heat–sensitive materials via thermal plasma processes. Nanoparticle sizes and interparticle spacing can be adjusted to enable specific electric charge transport and resistivity mechanisms. Exposure to low–temperature plasma has proved useful to improve the nucleation density, size and position uniformity of the self–organized metal nanoarrays; these possibilities have been predicted numerically and demonstrated experimentally. Gold nanoparticles using non–thermal plasma have been prepared by the synthesis of faceted germanium nanocrystals in a non–thermal plasma approach has also been demonstrated. A morphology evolution with particle size was observed for germanium crystals. Small particles were mainly of cubic shape with (100) facets, intermediate particles were cuboctahedrons, and large particles were spheres.

A collection of TEM images, illustrates the essential characteristics of the Ag particles which were grown with the FBMW method. Given the high metal loadings ($\sim$8 wt%) and the high energetic conditions of the plasma, it may be expected the metal particles would migrate and coalesce, however, the TEM observation does not suggest an occurrence of that phenomenon; the TEM observation reveals that Ag particles can coexist in very close proximity, even under the strong electronic TEM radiation for long periods of time. Thus, the Ag particles should be very stable. Both precursors yielded almost identical results in terms of nanoparticle density and shape. In summary, we have presented a method for preparing silica–alumina supported silver nanoparticles by combining two prevailing technologies: (1) microwave plasma inorganic synthesis; and (2) fluidized bed reactor. While the microwave radiation is the driving force in the production of the silver nanoparticles, the chaotic behavior of the fluidized bed homogenizes the particle distribution along the substrate surface. The plasma energy is punctually directed to the chemical species, transforming the silver precursors into metallic nanoparticles while damage to the support is minimized. The technique provides a relatively easy method to grow stable nanoparticles with an average 13 nm diameter in this case. We demonstrate the performance of the Ag/silica–alumina composite for the HDS reaction, which is similar to that of MoS_2 based catalysts. The bactericidal effect was also shown to be substantial against E. coli. Compared to related methods to produce supported nanoparticles, the inherent simplicity of the FBMW method is its main advantage, and for this reason, it could be easily scaled–up for industrial production. The nano–Ag/SiO_2–Al_2O_3 metal–ceramic composite can specifically be used as either a catalyst because it is active in the hydrodesulphurization of dibenzothiophene or as an antibacterial sand-filter due to its bactericidal effect on E. coli.

8.4.3 Synthesis of Nanoparticles

Modifying particle surface property with a special functional group can provide many useful applications in the fields of biomedical, cosmeceutical, pharmaceutical and polymerization materials. There are two basic approaches for surface modification: wet methods and dry methods such as plasma or radiation treatment (Lu et al. 2014a, b).

The former requires additional processes such as washing and drying, and yields an inhomogeneous modified surface. The surface modification by plasma treatment is a clean process that modifies the surface of the material without any damage within a short reaction time (Lu et al. 2014a, b; Du 2014a, c). However, most of the developed reactors for plasma treatment to date are suitable only for films (Bretagnol et al. 2004) and few reactors are capable for modifying a powder surface (Lu et al. 2014a, b; Du et al. 2014c, 2015a, b; Du 2014a, b). The reactors for powder treatment can be classified into three types: static bed (Du et al. 2015a, b; Du 2014a), moving bed (Lu et al. 2014a, b), and fluidized bed (Du 2014b, f; Du et al. 2013) reactors. Among these reactors, the fluidized bed has shown the particular features since powders are suspended while the plasma is generated in the upper part of the reactor. This method also has many advantages including homogeneous mixing between powders and the reactive gas, easily controlled continuous operation, maintaining isothermal condition throughout the reactor, and the amenability to large scale operation.

Polyethylene glycol (PEG) has non–toxic, odorless, neutral, nonvolatile, non-irritating and water–soluble properties that have led to use of PEG in various pharmaceutical products (El–Naas et al. 1998; Allah and Whitehead 2015). Moreover, PEG is well–known as its extraordinary ability to resist protein adsorption resulting from its hydrophilicity, large exclude volume and unique coordination with surrounding water molecules in aqueous mole (Allah and Whitehead 2015; Mutel et al. 2003; Leroy et al. 2003) wever, most of the developed reactors for plasma treatment to date are suitable only for films (Du et al. 2015a, b) and few reactors are capable for modifying a powder surface (Lu et al. 2014a, b; Du et al. 2013, 2014c, 2015a, b; Du 2014a, b, f). The reactors for powder treatment can be classified into three types: static bed (Weinberg et al. 1988; Flamant 1990), moving bed (Francke and Amouroux 1997), and fluidized bed (Sabat et al. 2014, 2015; Hu et al. 2015; Dasan et al. 2016; Li et al. 2014; Han et al. 2014; Chen et al. 2014; Tang et al. 2013; Ren et al. 2013; Roth et al. 2012; Baskakov 1964; Goldberger 1968; Sarjeant and ROY 1967; Sachs and Wirth 2015; Manieh et al. 1974; Kreibaum 1986) reactors. Among these reactors, the fluidized bed has shown the particular features since powders are suspended while the plasma is generated in the upper part of the reactor. This method also has many advantages including homogeneous mixing between powders and the reactive gas, easily controlled continuous operation, maintaining isothermal condition throughout the reactor, and the amenability to large scale operation.

Polyethylene glycol (PEG) has non–toxic, odorless, neutral, nonvolatile, non-irritating and water–soluble properties that have led to use of PEG in various pharmaceutical products (Du 2014a). Moreover, PEG is well–known as its extraordinary ability to resist protein adsorption resulting from its hydrophilicity, large exclude volume and unique coordination with surrounding water molecules in aqueous mole (Chen et al. 2006; Weinberg et al. 1988; Flamant 1990; Matsukata et al. 1992; Harker and Goldberg 1994; Francke and Amouroux 1997; Lin and Chyang 2003; Byun et al. 2011).

FB technology for nano–material synthesis offers the advantageous potential of high production yields (Du et al. 2008).

References

Allah ZA, Whitehead JC. Plasma–catalytic dry reforming of methane in an atmospheric pressure AC gliding arc discharge. Catal Today. 2015;256:76–9.

Al–Shamery K, Horowitz G, Sitter H, Rybahn HG. Interface controlled organic thin films. Berlin: Springer; 2009.

Aranovich BS, Peliks AA, Gluz MD, Borodulya VS, Ganzha VL. Graphite entrainment in reactors with electrothermal fluidized bed. Fibre Chem. 1973;4(5):540–2.

Arnauld P, Cavadias S, Amouroux J. The interaction of a fluidized bed with a thermal plasma: application to limestone decomposition. In: 7th international symposium on plasma chemistry (Eindhoven, 1985); 1985.

Arpagaus C, Rossi A, Rohr PRV. Short–time plasma surface modification of HDPE powder in a plasma downer reactor–process, wettability improvement and ageing effects. Appl Surf Sci. 2005;252(5):1581–95.

Attri P, Arora B, Choi EH. Utility of plasma: a new road from physics to chemistry. RSC Adv. 2013;3(31):12540–67.

Bai XY, Yi WM, Wang LH, Li YJ, Cai HZ. Fast pyrolysis of corn stalk for bio-oil in a plasma heated fluidized bed. Trans CSAE. 2005;21(12):127–30.

Bao WR, Chang LP, Lu YK. Study on main factors influencing acetylene formation during coal pyrolysis in arc plasma. Process Saf Environ. 2006;84(3):222–6.

Bartolomeu R, Foix M, Fernandes A, Tatoulian M, Ribeiro MF, Henriques C, et al. Fluidized bed plasma for pre-treatment of Co-ferrierite catalysts: an approach to NOx abatement. Catal Today. 2011;176(1):234–8.

Bashlai KI, Barantsev IF, Grinbaum MB, Stanyakin VM, Samodurov VV, Todes OM. Thermal and electrical characteristics of a high-frequency electrothermal fluidization bed. J Eng Phys. 1972;22(6):665–9.

Baskakov AP. The mechanism of heat transfer between a fluidized bed and a surface. Int J Chem React Eng. 1964;4:320–4.

Bieri J, Bialczak W, Wystalska K. Solid waste vitrification using a direct current plasma arc. In: Environmental engineering-proceedings of the 2nd national congress on environmental engineering (Netherlands, 2007); 2007.

Bourdin E, Fauchais P, Boulos M. Transient heat conduction under plasma conditions. Int J Heat Mass Tran. 1983;26(4):567–82.

Bretagnol F, Tatoulian M, Arefi-Khonsari F, Lorang G, Amouroux J. Surface modification of polyethylene powder by nitrogen and ammonia low pressure plasma in a fluidized bed reactor. React Funct Polym. 2004;61(2):221–32.

Bullard DE, Lynch DC. Reduction of ilmenite in a nonequilibrium hydrogen plasma. Metall Mater Trans B. 1997a;28(3):517–9.

Bullard DE, Lynch DC. Reduction of titanium dioxide in a nonequilibrium hydrogen plasma. Metall Mater Trans B. 1997b;28(6):1069–80.

Butscher D, Schlup T, Roth C, Müller-Fischer N, Gantenbein-Demarchi C, Rohr PRV. Inactivation of microorganisms on granular materials: reduction of *Bacillus amyloliquefaciens* endospores on wheat grains in a low pressure plasma circulating fluidized bed reactor. J Food Eng. 2015;159:48–56.

Byun Y, Cho M, Chung JW, Namkung W, Lee HD, Jang SD, et al. Hydrogen recovery from the thermal plasma gasification of solid waste. J Hazard Mater. 2011;190(1–3):317–23.

Chen X, Chen J, Wang Y. Unsteady heating of metallic particles in a rarefied plasma. Plasma Chem Plasma P. 1995;15(2):199–219.

Chen X, Chyou YP, Lee YC, Pfender E. Heat transfer to a particle under plasma conditions with vapor contamination from the particle. Plasma Chem Plasma P. 1985;5(2):119–41.

Chen X, He P. Heat transfer from a rarefied plasma flow to a metallic or nonmetallic particle. Plasma Chem Plasma P. 1986;6(4):313–33.

Chen X, Pfender E. Unsteady heating and radiation effects of small particles in a thermal plasma. Plasma Chem Plasma P. 1982;2(3):293–316.

Chen X, Pfender E. Behavior of small particles in a thermal plasma flow. Plasma Chem Plasma P. 1983a;3(3):351–66.

Chen X, Pfender E. Effect of the Knudsen number on heat transfer to a particle immersed into a thermal plasma. Plasma Chem Plasma P. 1983b;3(1):97–113.

Chen GL, Fan SH, Li CL, Gu WC, Feng WR, Zhang GL, et al. A novel atmospheric pressure plasma fluidized bed and its application in mutation of plant seeds. Chin Phys Lett. 2005;22 (8):1980–3.

Chen G, Chen S, Zhou M, Feng W, Gu W, Yang S. Application of a novel atmospheric pressure plasma fluidized bed in the powder surface modification. J Phys D Appl Phys. 2006;39 (24):5211.

Chen J, Cheng Y, Xiong X, Wu C, Jin Y. Research progress of coal pyrolysis to acetylene in thermal plasma reactor. Chem Ind Eng Prog. 2009;28(3):361–7.

Chen Z, Dai XJ, Magniez K, Lamb PR, Fox BL, Wang X. Improving the mechanical properties of multiwalled carbon nanotube/epoxy nanocomposites using polymerization in a stirring plasma system. Compos Part A-Appl S. 2014;56(56):172–80.

Cormier JM, Rusu I. Syngas production via methane steam reforming with oxygen: plasma reactors versus chemical reactors. J Phys D Appl Phys. 2001;34(34):2798.

Currier R, Blacic J. Plasma processing of lunar and planetary materials. In: Space resources roundtable II (Colorado, 2000); 2000.

Currier R, Trkula M. Hydrogen plasma reduction of planetary materials. In: ISRU III technical interchange meeting (Denver, 1999); 2000.

Dasan BG, Mutlu M, Boyaci IH. Decontamination of *Aspergillus flavus* and *Aspergillus parasiticus* spores on hazelnuts via atmospheric pressure fluidized bed plasma reactor. Int J Food Microbiol. 2016;216:50–9.

Du CM. A plasma fluidized bed for the production of syngas from MSW. China patent 201410844203.9; 2014a.

Du CM. A plasma fluidized bed for the cineration of fly ash. China patent 201410850031.6; 2014b.

Du CM. A plasma fluidized bed for abatement of VOCs. China patent 201410850032.0; 2014c.

Du CM. A plasma–catalytic fluidized bed for VOC abatement. China patent 201410849939.5; 2014d.

Du CM. The remediation of organic contamimated soil by a plasma fluidized bed. China patent 201410849940.8; 2014f.

Du CM, Xiao MD. Cu_2O nanoparticles synthesis by microplasma. Sci Rep-UK. 2014;4:7339.

Du CM, Yan JH. Electrical and spectral characteristics of a hybrid gliding arc discharge in air–water. IEEE T Plasma Sci. 2007;35(6):1648–50.

Du CM, Shi TH, Sun Y, Zhuang X. Decolorization of acid orange 7 solution by gas–liquid gliding arc discharge plasma. J Hazard Mater. 2008;154(1–3):1192.

Du CM, Zhang LL, Wang J, Zhang CR, Li HX, Xiong Y. Degradation of acid orange 7 by gliding arc discharge plasma in combination with advanced fenton catalysis. Plasma Chem Plasma P. 2010;30(6):855–71.

Du CM, Li H, Zhang L, Wang J, Huang D, Xiao M, et al. Hydrogen production by steam-oxidative reforming of bio-ethanol assisted by Laval nozzle arc discharge. Int J Hydrogen Energ. 2012;37(10):8318–29.

Du CM, Huang DW, Li HX, Xiao MD, Wang K, Zhang L, et al. Adsorption of acid orange II from aqueous solution by plasma modified activated carbon fibers. Plasma Chem Plasma P. 2013;33 (1):65–82.

Du CM, Huang DW, Mo JM, Ma DY, Wang QK, Mo ZX, et al. Renewable hydrogen from ethanol by a miniaturized nonthermal arc plasma-catalytic reforming system. Int J Hydrogen Energ. 2014a;39(17):9057–69.

Du CM, Tang J, Mo JM, Ma DY, Wang J, Wang K, et al. Decontamination of bacteria by gas-liquid gliding arc discharge: application to. IEEE T Plasma Sci. 2014b;42(9):2221–8.

Du CM, Ma DY, Wu J, Lin YC, Xiao W, Ruan JJ, et al. Plasma-catalysis reforming for H2 production from ethanol. Int J Hydrogen Energ. 2015a;40(45):15398–410.

Du CM, Wu J, Ma DY, Liu Y, Qiu PP, Qiu RL, et al. Gasification of corn cob using non-thermal arc plasma. Int J Hydrogen Energ. 2015b;40(37):12634–49.

El–Naas MH. Synthesis of calcium carbide in a plasma spout fluid bed. Canada: McGill University; 1996.

El–Naas MH, Munz R, Ajersch F. Solid–phase synthesis of calcium carbide in a plasma reactor. Plasma Chem Plasma P. 1998;18(3):409–27.

Emome A, Jurewize T. Fuel synthesis for solid oxide fuel cells by plasma spouted bed gasification. In: 14th international symposium on plasma chemistry (Prague, 1999); 1999.

Fabry F, Rehmet C, Rohani V, Fulcheri L. Waste gasification by thermal plasma: a review. Waste Biomass Valori. 2013;4(3):421–39.

Flamant G. Hydrodynamics and heat transfer in a plasma spouted bed reactor. Plasma Chem Plasma P. 1990;110(1):71–85.

Flamant G. Plasma fluidized and spouted bed reactors: an overview. Pure Appl Chem. 2009;66(6):1231–8.

Francke E, Amouroux J. LDA simultaneous measurements of local density and velocity distribution of particles in plasma fluidized bed at atmospheric pressure. Plasma Chem Plasma P. 1997;17(4):433–52.

Goldberger WM. Method of generating a plasma arc with a fluidized bed as one electrode. United States Patent 3404078; 1968.

Goldberger WM, Oxley JH. Quenching the plasma reaction by means of the fluidized bed. AIChE J. 1963;9(6):778–82.

Gomez E, Rani DA, Cheeseman CR, Deegan D, Wise M, Boccaccini AR. Thermal plasma technology for the treatment of wastes: a critical review. J Hazard Mater. 2008;161(2–3):614–26.

Grovender EA, Cooney CL, Langer RS, Ameer GA. Modeling the mixing behavior of a novel fluidized extracorporeal immunoadsorber. Chem Eng Sci. 2001;56(18):5437–41.

Han SU, Na YH, Yong CH, Dong HS, Chang HC, Park YK. High-efficiency gasification of low-grade coal by microwave steam plasma. Energ Fuel. 2014;28(7):4402–8.

Harker AB, Goldberg IB. Diamond growth by microwave generated plasma flame. United States Patent 5349154; 1994.

Heberlein J, Murphy AB. Thermal plasma waste treatment. J Phys D Appl Phys. 2008;41(5):053001.

Hu MB, Dang SC, Ma Q, Xia WD. Stabilizing effect of plasma discharge on bubbling fluidized granular bed. Chinese Phys B. 2015;24(7):288–92.

Jung SH, Park SM, Park SH, Kim SD. Surface modification of fine powders by atmospheric pressure plasma in a circulating fluidized bed reactor. Ind Eng Chem Res. 2004;43(18):5483–8.

Karches M, Bayer C, Rohr PRV. A circulating fluidised bed for plasma-enhanced Chem Vapor Depos on powders at low temperatures. Surf Coat Tech. 1999;116–119(4):879–85.

Kim GH, Kim SD, Park SH. Plasma enhanced chemical vapor deposition of TiO_2 films on silica gel powders at atmospheric pressure in a circulating fluidized bed reactor. Chem Eng Process. 2009;48(6):1135–9.

Kogelschatz U, Eliasson B, Egli W. Dielectric-barrier discharges. Principle and applications. J Phys IV. 1997;07(C4):47–66.

Krawczyk K, Ulejczyk B, Song H, Lamenta A, Paluch B, Schmidt–Szalowski K. Plasma-catalytic reactor for decomposition of chlorinated hydrocarbons. Plasma Chem Plasma P. 2009;29(1):27–41.

Kreibaum J. Plasma spouted bed calcination of lac Doré vanadium ore concentrate. Canada: McGill University; 1986.

Kumar A, Dwivedi HK, Nehra V. Atmospheric non-thermal plasma sources. Int J Eng. 2008;2(1):53–68.

Laroussi M. Low-temperature plasmas for medicine? IEEE T Plasma Sci. 2009;37(6):714–25.

Leroy JB, Fatah N, Mutel B, Grimblot J. Treatment of a polyethylene powder using a remote nitrogen plasma reactor coupled with a fluidized bed: influence on wettability and flowability. Plasmas Polym. 2003;8(1):13–29.

Lesinski JMBJ, Meillot E, Debbagh–Nour G. Modelling of plasma entrianded bed coal gasifiers. In: 7th international symposium on plasma chemistry (Eindhoven, 1985); 1985.

Li MW, Gonzalez-Aguilar J, Fulcheri L. Synthesis of titania nanoparticles using a compact nonequilibrium plasma torch. Jpn J Appl Phys. 2008;47(9):7343–5.

Li SD, Tian SH, Du CM, He C, Cen CP, Xiong Y. Vaseline–loaded expanded graphite as a new adsorbent for toluene. Chem Eng J. 2010;162(2):546–51.

Li X, Han J, Wu CN, Guo Y, Yan B, Cheng Y. Coal tar pyrolysis to acetylene in thermal plasma. CIESC Journal. 2014;65(9):3680–6.

Lin YC, Chyang CS. Radial gas mixing in a fluidized bed using response surface methodology. Powder Technol. 2003;131(1):48–55.

Liu LX, Rudolph V, Litster J. A direct current, plasma fluidized bed reactor: its characteristics and application in diamond synthesis. Powder Technol. 1996;88(1):65–70.

Lu S, Chen L, Huang Q, Yang L, Du C, Li X, et al. Decomposition of ammonia and hydrogen sulfide in simulated sludge drying waste gas by a novel non-thermal plasma. Chemosphere. 2014a;117:781–5.

Lu SY, Chen L, Huang QX, Yang LQ, Du CM, Li XD, et al. Decomposition of ammonia and hydrogen sulfide in simulated sludge drying waste gas by a novel non-thermal plasma. Chemosphere. 2014b;117(117C):781.

Lundholm K, Nordin A, Öhman M, Boström D. Reduced bed agglomeration by co-combustion biomass with peat fuels in a fluidized bed. Energ Fuel. 2005;19(6):2273–8.

Ma JC, Zhao HB, Guo L, Zheng CG. Investigations on batch preparation of iron-based oxygen carrier by spouted bed and using in chemical looping combustion of coal. J Eng Thermophys. 2013;34(10):1960–3.

Manieh AA, Scott DS, Spink DR. Electrothermal fluidized bed chlorination of zircon. Can J Chem Eng. 1974;52(4):507–14.

Matsukata M, Oh–hashi H, Kojima T, Mitsuyoshi Y, Ueyama K. Vertical progress of methane conversion in a DC plasma fluidized bed reactor. Chem Eng Sci. 1992;47(9–11):2963–8.

Mendes A, Dollet A, Ablitzer C, Perrais C, Flamant G. Numerical simulation of reactive transfers in spouted beds at high temperature: application to coal gasification. J Anal Appl Pyrol. 2008;82(1):117–28.

Mochizuki Y, Ono S, Teii S, Chang J. Fluidization and plasma characteristics of medium pressure RF glow discharge plasma fluidized bed reactors. Adv Powder Technol. 1993;4(3):159–67.

Moissan H, Dewar J. Nouvelles expériences sur la liquéfaction de fluor. C R. 1897;125:505–11.

Morstein M, Karches M, Bayer C, Casanova D, Rohr PRV. Plasma CVD of ultrathin TiO2 films on powders in a circulating fluidized bed. Chem Vapor Depos. 2000;6(1):16–20.

Mutel B, Fatah N, Vezin H, Grimblot J. Surface Modification of polyethylene powders using a remote nitrogen plasma and a fluidized bed reactor. In: 16th international symposium on plasma chemistry Taormina (Italy, 2003); 2003.

Nessim C, Boulos M, Kogelschatz U. In–flight coating of nanoparticles in atmospheric–pressure DBD torch plasmas. EUR Phys J-Appl Phys. 2009;47(2):22819.

Nezu A, Morishima T, Watanabe T. Thermal plasma treatment of waste ion-exchange resins doped with metals. Thin Solid Films. 2003;435(1):335–9.

Nikravech M, Baba K, Lazzaroni C. Development of fluidized spray plasma (FSP) device to deposit nanostructural catalysts on ceramic beads. In: 22nd international symposium on plasma chemistry (Antwerp, 2015); 2015.

Pajkic Z, Willert-Porada M. Arc-PVD coating of powders in a microwave plasma fluidized bed. In: IEEE 35th international conference on plasma science (Karlsruhe, 2008); 2008.

Pajkic Z, Willert-Porada M. Atmospheric pressure microwave plasma fluidized bed CVD of AlN coatings. Surf Coat Tech. 2009;203(20):3168–72.

Park SM, Jung SH, Park SH, Kim SD. Silicon oxide thin film deposition on alumina in a circulating fluidized bed reactor. Key Eng Mat. 2005;277–279:577–82.

Pfender E. Thermal plasma technology: where do we stand and where are we going? Plasma Chem Plasma P. 1999;19(1):1–31.

Ren Y, Li XD, Yu L, Cheng K, Yan JH, Du CM. Degradation of PCDD/Fs in fly ash by vortex–shaped gliding arc plasma. Plasma Chem Plasma P. 2013;33(1):293–305.

Renzo AD, Maio FPD. Homogeneous and bubbling fluidization regimes in DEM–CFD simulations: hydrodynamic stability of gas and liquid fluidized beds. Chem Eng Sci. 2007;62(1–2):116–30.

Rohr PRV, Borer B. Plasma–enhanced CVD for particle synthesis using circulating fluidized bed technology. Chem Vapor Depos. 2007;13(9):499–506.

Roth C, Keller L, Rohr PRV. Adjusting dissolution time and flowability of salicylic acid powder in a two stage plasma process. Surf Coat Tech. 2012;206(19–20):3832–8.

Sabat KC, Rajput P, Paramguru RK, Bhoi B, Mishra BK. Reduction of oxide minerals by hydrogen plasma: an overview. Plasma Chem Plasma P. 2014;34(1):1–23.

Sabat KC, Paramguru RK, Pradhan S, Mishra BK. Reduction of cobalt oxide (Co_3O_4) by low temperature hydrogen plasma. Plasma Chem Plasma P. 2015;35(2):387–99.

Sachs M, Wirth K. Effect of oxygen content on the functionalization of polymer powders using an atmospheric plasma jet in combination with a fluidized bed reactor. In: 22nd international symposium on plasma chemistry (Antwerp, 2015); 2015.

Sakano M, Tanaka M, Watanabe T. Application of radio–frequency thermal plasmas to treatment of fly ash. Thin Solid Films. 2001;386(2):189–94.

Sanchez I, Flamant G, Gauthier D, Flamand R, Badie JM, Mazza G. Plasma-enhanced chemical vapor deposition of nitrides on fluidized particles. Powder Technol. 2001;120(1–2):134–40.

Santoianni J, Heier ME, Gorodetsky A, Reese S, Hicks KO. Plasma gasification reacotrs with modified carbon beds and reduced coke requirement. United States Patent 9222026; 2015.

Sarjeant PT, Roy R. New glassy and polymorphic oxide phases using rapid quenching techniques. J Am Ceram Soc. 1967;50(10):500–3.

Sathiyamoorthy D. Plasma spouted/fluidized bed for materials processing. J Phys Conf Ser. 2010;012120.

Savintsev MI. Diffusion saturation in electrothermal fluidized bed. Met Sci Heat Treat. 1990;32 (11):842–5.

Schmidt-Szalowski K, Krawczyck K, Mlotek M. Properties of a heterogeneous system of solid particles in gliding discharge plasma. In: 10th international symposium on high pressure low temperature plasma chemistry (Hakone X, 2006); 2006.

Şen Y, Bağcı U, Güleç HA, Mutlu M. Modification of food-contacting surfaces by plasma polymerization technique: reducing the biofouling of microorganisms on stainless steel surface. Food Bioprocess Tech. 2012;5(1):166–75.

Shi TH, Jia SG, Chen Y, Wen YH, Du CM, Guo HL, et al. Adsorption of Pb (II), Cr (III), Cu (II), Cd (II) and Ni (II) onto a vanadium mine tailing from aqueous solution. J Hazard Mater. 2009a;168(1–3):838.

Shi TH, Wang ZC, Liu Y, Jia SG, Du CM. Removal of hexavalent chromium from aqueous solutions by D301, D314 and D354 anion–exchange resins. J Hazard Mater. 2009b;161(2–3):900–6.

Song LH, Park SH, Jung SH, Sang DK, Park SB. Synthesis of polyethylene glycol-polystyrene core-shell structure particles in a plasma-fluidized bed reactor. Korean J Chem Eng. 2011;28 (2):627–32.

Spillmann A, Sonnenfeld A, Rohr PRV. Flowability modification of lactose powder by plasma enhanced Chem Vapor Depos. Plasma Process Polym. 2007;4(Supplement S1):S16–S20.

Tang L, Huang H, Hao H, Zhao K. Development of plasma pyrolysis/gasification systems for energy efficient and environmentally sound waste disposal. J Electrostat. 2013;71(5):839–47.

Taylor PR, Pirzada SA. Thermal plasma processing of materials: a review. Adv Perform Mater. 1994;1(1):35–50.

Thorley B, Saunby JB, Mathur KB, Osberg GL. An analysis of air and solid flow in a spouted wheat bed. Can J Chem Eng. 1959;37(5):184–92.

Tsukada M, Goto K, Yamamoto RH, Horio M. Metal powder granulation in a plasma-spouted/ fluidized bed. Powder Technol. 1995;82(3):347–53.

Uemura Y, Baba K, Ohe H, Ohzuno Y, Hatate Y. Catalytic decomposition of hydrocarbon into hydrogen and carbon in a spouted-bed reactor as the second–stage reactor of a plastic recycling process. J Mater Cycles Waste. 2003;5(2):94–7.

Uglov AA, Gnedovets AG. Effect of particle charging on momentum and heat transfer from rarefied plasma flow. Plasma Chem Plasma P. 1991;11(2):251–67.

Upadhya K, Moore JJ, Reid KJ. Application of thermodynamic and kinetic principles in the reduction of metal oxides by carbon in a plasma environment. Metall Trans B. 1986;17 (1):197–201.

Vivien C, Wartelle C, Mutel B, Grimblot J. Surface property modification of a polyethylene powder by coupling fluidized bed and far cold remote nitrogen plasma technologies. Surf Interface Anal. 2002;34(1):575–9.

Wang LH, Bai XY, Yi YM, Kong FX, Li YJ, He F, et al. Characteristics of bio-oil from plasma heated fluidized bed pyrolysis of corn stalk. Trans CSAE. 2006;663–665(3):502–5.

Wang Q, Cheng Y, Jin Y. Dry reforming of methane in an atmospheric pressure plasma fluidized bed with $Ni/\gamma-Al_2O_3$ catalyst. Catal Today. 2009;148(3):275–82.

Wang TC, Lu N, Li J, Wu Y. Degradation of pentachlorophenol in soil by pulsed corona discharge plasma. J Hazard Mater. 2010;180(1–3):436–41.

Wang TC, Lu N, Li J, Wu Y. Plasma-TiO_2 catalytic method for high-efficiency remediation of p-nitrophenol contaminated soil in pulsed discharge. Environ Sci Technol. 2011;45(21):9301–7.

Weinberg FJ, Bartleet TG, Carleton FB, Rimbotti P, BrophyJH, Manning R. Partial oxidation of fuel-rich mixtures in a spouted bed combustor. Combust Flame. 1988;72(3):235–9.

Wu CN, Yan BH, Zhang L, Shuang Y, Jin YT, Cheng Y. Analysis of key techniques and economic feasibility for one-step production of acetylene by coal pyrolysis in thermal plasma reactor. CIESC J; 2010.

Yamamoto T. VOC decomposition by nonthermal plasma processing—A new approach. J Electrostat. 1997;42(1–2):227–38.

Yan B, Xu P, Guo CY, Jin Y, Cheng Y. Experimental study on coal pyrolysis to acetylene in thermal plasma reactors. Chem Eng J. 2012;207–208(10):109–16.

Yan B, Cheng Y, Jin Y. Cross-scale modeling and simulation of coal pyrolysis to acetylene in hydrogen plasma reactors. AIChE J. 2013;59(6):2119–33.

Yang JS, Bao WR, Zhang YF, Xie KC. Engineering application study of producing acetylene through coal pyrolysis in plasma reactor. Chem Eng. 2006;34(6):52–5.

Ye QZ, Li J, Xie ZH. Analytical model of the breakdown mechanism in a two–phase mixture. J Phys D Appl Phys. 2004;37(24):3373.

Zhu CW, Zhao GY, Hlavacek V. A dc plasma–fluidized bed reactor for the production of calcium carbide. J Mater Sci. 1995;30(9):2412–9.

Zhu F, Zhang J, Yang Z, Guo Y, Li H, Zhang Y. The dispersion study of TiO_2 nanoparticles surface modified through plasma polymerization. Physica E. 2005;27(4):457–61.

Chapter 9
Comparison of the Performance with Different Plasma Fluidized Beds

Abstract In this chapter, the research and practical applications of plasma fluidized bed were made a comprehensive comparison from 1985 to now. The comparison of operating conditions including plasma system, plasma source and power input, temperature, carrier gas and flow rate, pressure and materials, but also lists their main products and applications, and highlights their own characteristics. It is suggested that the plasma fluidized bed can be used in different fields and has broad prospects, which deserves further study.

Keywords Comprehensive comparison · Applications · Plasma fluidized bed

Combination of plasma and fluidized bed is a relatively new and interdisciplinary field of research and technology. The present review has shown that, within this field, fluidized beds are a convenient medium to efficiently and economically utilize the energy injected by plasma. Indeed, the most conventional cases, precursor conversion and deposit uniformity are excellent, design, build-up and scale-up are easy, and equipment costs quite low. However, different plasma fluidized bed generally showed significant differences in respect to system design, operation mode as well as the potential application field.

To illustrate, when taking different plasma source, there existed significant differences between thermal plasma fluidized bed and non-thermal plasma fluidized bed. To begin with, the average temperature and the temperature distribution are much different, which has been mentioned in previous section. The average temperature inside thermal plasma fluidized bed is generally higher than 373 K, especially for plasma torch fluidized bed, the temperature is generally higher than 1000 K, which in some ways leads to the different application field and purpose of the thermal plasma fluidized bed and nonthermal plasma fluidized bed. More thermal plasma fluidized beds especially dc plasma torch fluidized beds have been used for high temperature application, such as metallurgy extraction or the pyrolysis

C. Du et al., *Plasma Fluidized Bed*, Advanced Topics in Science and Technology in China, https://doi.org/10.1007/978-981-10-5819-6_9

of the solid fuel or hydrocarbons, while more nonthermal plasma fluidized beds have been used for the generation of the large number of active species, such as radicals and excited species, and for the reaction with the active species, such as the utilization of the active species for the surface functionalization and very high productivity can be achieved with a high energy efficiency. Also, for high temperature operation, the more used carrier gases are hydrogen or hydrogen mixed with other types of gas, such as valuable gas, while for low temperature process, the carrier gas of the plasma fluidized bed are much more abundant in types and the ways of mixing.

On the other hand, various fluidized bed system can also cause distinguishes in many aspects, ranging from the operation condition to the heat transfer, hydrodynamic, the gas-solid contact, etc. Size of bed particles injected into the circulating plasma fluidized bed can be much smaller than other kinds of plasma fluidized bed, particularly, than that of spouted bed, which can be fitted to coarse particles whose diameter up to mm range. Secondly, inside different plasma fluidized bed, the forcing drafts are different, i.e. gravity and the carrier gas, which can significantly lead to the resident time and then the reaction time of the solid particles, for example, under the same condition, the resident time of the solid particles in the circulating fluidized bed can be much longer than that of the downer bed, which in turn affect the working efficiency of the system or the product selectivity and also the yield in unit time. Then, different fluidized system can also results in a different heat transfer and mass transfer efficiency, which can be summarized in the following order generally: fluidized bed > circulating fluidized bed > spouted bed > downer bed. Moreover, the temperature gradients inside the different plasma fluidized bed system also show various trends, obviously. The temperature distributions inside the plasma circulating fluidized bed and common fluidized bed are both quiet uniform and the spouted bed otherwise, which leads to the hot spot in the reaction zone.

All these differences due to the fluidized system can finally contribute to the application field and the related process flexibility. Therefore, spouted beds are much more benefit for high temperature process of the relatively coarse particles, while plasma circulating fluidized beds are much more suitable for surface modification of the particles for excellent features and downer bed for super short time process such as coal pyrolysis, showing great application potential, respectively.

The applications of plasma fluidized bed in metallurgy, energy, environmental protection and new material technology are compared by author, as shown in Table 9.1.

Table 9.1 Comparisons between various plasma fluidized bed

Researchers (country)	Year	Operation conditions			Main results		Highlights	Ref.
		System	Plasma source and power input; temperature(K)	Carrier gas; flow rate; pressure;	Materials; particle feature (solid mass)	Application	Main products		
Arnould et al. (France)	1985	Fluidized bed	RF (3 kW) 773–1373 K	Ar+N_2+Air; 2.5–3 ms^{-1}	Limestone in the range of 250 and 350 μm	Limestone decomposition	CaO	Good thermal transfer	(Arnould et al. 1985)
Mersereau (Canada)	1990	Spout fluid bed	DC plasma torch (20 kW); 633–1148 K	Ar+N_2; 110L/min; Atm	Salt and ore with diameter of 500 μm (100 g)	Calcination of lac dore vanadium	Sodium metavanadate	Complete conversion of the feed materials	(Mersereu 1990)
Zhu et al. (USA)	1995	Fluidized bed	DC arc torch 1673–1773 K; Argon; 0.7m^{3}h^{-1}		CaO+C with diameter larger than 2 mm	Production of calcium carbide	Calcium carbide with diameter less than 0.2 mm	Low process temperature; clean products and less material loss	(Zhu et al. 1995)
Tsukada et al.	1995	Spouted fluidized bed	DC plasma torch (900 W) 450–550 K	Argon+H_2; 3L/min; Atm	Iron powder (149–210 um) and aluminum powder (74–88 um)	Metal powder granulation	Spherical ally grain		(Tsukada et al. 1995)
M.H.El-Nass (Canada)	1997	Fluidized bed	Dc plasma torch; (20 kW); 2150 K	Argon; 40L/min Atm	Calcium oxide with 170 μm and graphite with 130 μm (1 kg)	Synthesis of calcium carbide	Calcium carbide	Lower the energy consumption up to 40%	(El-Naas 1996)
M.H.El-Nass et al. (Canada)	1998	Spout-fluidized bed	Dc plasma torch (20 kW); 1573 K	Argon; 40L/min	Calcium oxide with 170 μm and graphite with 130 μm (1 kg)	Synhesis of calcium carbide	Calcium carbide	Simultaneous synthesis and granulation	(El-Naas et al. 1998)
Currier et al. (USA)	1999	Downer bed	Microwave discharge	Hydrogen	Anhydrous silicate and oxide minerals	Reduction of planetary materials	Ti and Al metal and oxygen	Good gas-solid contact and great potential in low gravity atmosphere	(Currier et al. 1999)
Currier and Blacic (USA)	2000	Downer bed	Microwave discharge; hydrogen		Ilmenite powder	Reduction of lunar and planetary materials	Oxygen and Ti	Simultaneous extractive metallurgy and oxygen recovery	(Currier and Blacic 2000)
Liang et al. (China)	2009	Fluidized bed	Microwave discharge (20 kW) 773 K	Air+H_2O; 0.4–0.8 m^3/h	High carbon ferromanganese powder (500 g)	Solid phase decarbonization of high carbon ferromanganese powder	Solid phase decarbonization of high carbon ferromanganese powder	free from external contamination for materials	(Liang et al. 2009)

(continued)

Table 9.1 (continued)

Researchers (country)	Year	Operation conditions		Main results			Highlights	Ref.
		System	Plasma source and power input; temperature(K)	Carrier gas; flow rate; pressure;	Materials; particle feature (solid mass)	Application	Main products		
Claude Andre Lelievre (Canada)	1996	Fluidized bed	Single phase arc plasma; 600 °C; 200 + 950 °C	Argon+oxygen+CO_2; 2.45L/min; 1 torr					
Emome and Jurewize (Canada)	1999	Spouted bed	Plasma torch (9.4 kW); 983 K	Argon+CO_2; 2.5 plm; atmosphere pressure	Canola 1.5 mm	Biomass gasification	H_2+CH_4+C_2H_2+CO	Absence of tary by products and high overall energy efficiency and high selectivity of spouted fluidized bed	(Emome and Jurewize 1999)
Bai et al. (China)	2005	Fluidized bed	Dc plasma jet; 710–790 K	Argon; 0.89 m/s	Sand and Corn stalk powder	Biomass pyrolysis	Bio-oil		(Bai et al. 2005)
Wang et al. (China)	2006	Fluidized bed	Dc plasma jet; 710–790 K		Sand and corn stalk	Biomass pyrolysis	Bio-oil	Stable and easy operation; fast heating	(Wang et al. 2006)
BAO et al. (China)	2006	Downer bed	arc plasma jet; 1273 K	Hydrogen and oxygen; 2.0 Nm3/hand 1.0 Nm^3/h	Coal samples particle size of less than 0.074 mm (5 g s^{-1})	Coal pyrolysis	The acetylene and carbon monoxide were the main gaseous products	Simple and environmentally friendly	(Bao et al. 2006)
Schmidt-Szałowski et al.	2007	Spouted bed	GD (100–400 W)	0.4 CH_4+0.6 H_2 300 Nl/h; atmospheric pressure	Catalyst particles(0.16–0.315 mm)	Methane reformation	C_2 hydrocarbons	High product selectivity	(Schmidt-Szałowski et al. 2007)
Młotek et al. (Poland)	2009	Spout fluidized bed	GD; 413–573 K	0.4CH_4+0.6H_2; 1640 Nl/h; Atm	Alumina-ceramic supported catalysts (Pt/Al_2O_3 and Pd/Al_2O_3,) 0.16–0.315 mm	Methane reformation	Ethane and ethylene	Reduction of carbon (soot) formation and increase in selectivity to C_2 hydrocarbons	(Młotek et al. 2009)
Wang et al. (China)	2009	Fluidized bed	DBD (20–90 W); 648 K–798 K	CH_4+CO_2 5 ml/min to 50 ml/min; atm	Ni/γ–Al_2O_3 catalyst	Dry reforming of methane	C_2	new features to the synergetic effect of plasma and catalyst	(Wang et al. 2009)
Zhang et al. (China)	2009	Fluidized bed	Dc plasma torch;	Argon; 2.4–2.6 m^3/h	Sand(0.45–0.6 mm) and Corn stalk powder (< 0.6 mm)	Biomass pyrolysis	Bio-oil	Stable and easy operation; fast heating	(Zhang et al.2009)
Lee and Sekiguchi (Japan)	2011	Spouted bed	Gliding arc discharge (17 W); 443 K	Argon+CH_4 600 ml min^{-1} and 1000 ml min^{-1}	Alumina-supported catalysts with Pt, Pd, Rh, and Ru	CH_4 reformation	C_2H_2, H_2 and soot	High selectivity of products	(Lee and Sekiguchi 2011)

(continued)

Table 9.1 (continued)

Researchers (country)	Year	Operation conditions				Main results		Highlights	Ref.
		System	Plasma source and power input; temperature(K)	Carrier gas; flow rate; pressure;	Materials; particle feature (solid mass)	Application	Main products		
Kroker et al.	2012	Fluidized bed	RF (20–50 W); 250 °C	(60% methane and 40% carbon dioxide; 100 mbar	Alumina-ceramic supported catalysts containing Cu and Pd	CH_4 reformation	H_2/CO	The highest yield and selectivity of hydrogen is found with the fluidized Pd/Al_2O_3 Ca	(Kroker et al. 2012)
Yan et al. (China)	2012	Downer bed	DC plasma torch (1.5–3.0 kW); 2000 K	Ar/H_2; 16.7 L/min	Coal particles 10–100 um; 2 g/min	Plasma pyrolysis of coal	C_2H_2	Proposed an alternative, simple method for preliminary selection of coal rank	(Yan et al. 2012)
Yan et al. (China)	2013	Downer bed	V-shaped plasma torch (5–MW/2–MW); 1600 K	$Ar+H_2$; Atm	Coal particles with 20–50 μm; 700 or 1700 kg/h	Coal Pyrolysis to Acetylene	C_2H_2	Complete heating of particles is possible	(Yan et al. 2013)
LI et al. (China)	2014	Downer bed	Non-transferred arc plasma torch (8 kW)	80% Ar+20% H_2; 15 $L·min^{-1}$; 20 $g·min^{-1}$	Coal samples particle size	Coal tar pyrolysis	Acetylene	A new way to conversion the coal to high value products	(Li et al. 2014)
Mohammedi et al. (France)	1995	Fluidized bed	ICP with a 5.4 MHz (18 kW) 5000 K	$Ar+H_2$; 30L/min	CaO with 250–500 um	Hydrogen radical production	Hydrogen radical	High rates of heat and mass transfer such as an efficient quenching of highly reactive species and out of equilibrium gas	(Mohammedi et al. 1995)
Sekiguchi et al. (Japan)	1995	Fluidized bed	DC plasma torch (3.2 kW)	CH_3Cl+O_2+Ar; 10L/min	CaO particles (149–840 um) 50 g	Decomposition of chloromethane and recovery of hydrogen chloride	$CaCl_2$	Simple operation	(Sekiguchi et al. 1995)
M.Langlerron et al. (France)	2000	Fluidized bed	RF (75 W)	Oxygen+Toluene; 80 Pa	Alumina catalyst particles with a diameter of 350–500 um	Selective oxidation of toluene	CO and CO_2	Fully degradation	
Ramachandran and Kikukawa (Japan)	2000	Downer reactor	DC non-transferred arc (7–16 kW)	$Ar+H_2+N_2+O_2$	Powdered electroplating sludge (size 45 μm)	Treatment of electroplating sludge	Non-leachable slag	A new plasma recycle process for hazardous wastes, especially for electroplating sludge	(Ramachandran and Kikukawa 2000)

(continued)

Table 9.1 (continued)

Researchers (country)	Year	Operation conditions				Main results		Highlights	Ref.
		System	Plasma source and power input; temperature(K)	Carrier gas; flow rate; pressure;	Materials; particle feature (solid mass)	Application	Main products		
Steinbach et al. (USA)	2003	Fluidized bed	Electro-thermal plasma; > 1273 K	He 200–300 ml/min	Carbon particles	Reduction of small inorganic gases	Removal of oxygen)	A cleaner, burning effluent gas	(Steinbach et al. 2003)
Vaidyanathan et al. (USA)	2007	Downer reactor	Plasma arc torch (100 kW); 1200 K	Air; 0.19–0.28 m^3/s	Carpet waste and solid wastes generated by a USAF BEAR Base deployment	Treatment of two solid waste streams	Carbon monoxide and hydrogen	Non-polluting and capable of destroying substances harmful to human health, and the products are pre-dictable, harmless and acceptable for public health and environment	(Vaidyanathan et al. 2007)
Chen et al. (China)	2006	Fluidized bed	DBD (12 W); 315–320 K	Argon+Oxygen; 1Lmin^{-1}	Plant seed	Mutation of seed	Mutated comb seed and unmutated pimiento seed	Continuous process; without local burnt	(Chen et al. 2006)
Butscher et al. (Switzerland)	2015	Circulating fluidized bed	RF-generator (700–900 W)	8–12.8 mbar; argon with varying oxygen admixture;	Wheat grains 500 g of grains	Reduction of Bacillus amyloliquefaciens endospores on wheat grains		No negative effects of plasma treatment on flour and baking properties	(Butscher et al. 2015)
Karches et al. (Switzerland)	1999	Circulating fluidised bed	2.45 GHz microwave plasma (500 W); 368 K	Argon+oxygen+HMDSO; 10 ms^{-1} 400 Pa	NaCl crystals 210 um; 0–50 g l^{-1}	PECVD	NaCl crystals coated with a thin silicon oxide film	High homogeneity; low process temperature; high rate; short treatment time	(Karches et al. 1999)
Park and Kim (Korea)	1999	Fluidized bed	RF plasma (100 W)	CF_4+He(5+20sccm)	High density polyethylene powder with mean diameter of 231 μm	Surface functionalization	Fluorine content in the surface of HDPE powder	Good modification efficiency	(Park and Kim 1999)
Takarada et al. (Japan)	1993	Fluidized bed	RF 200 W 1193–1223 K	30 torr 500 cm/m^3	Single crystal silicon particle of 2000–4000 μm	Diamond like carbon coating	Fine diamond particle deposited on the non-diamond surfaces	Effective nucleation of diamond	(Takarada et al. 1993)
Liu et al. (Australia)	1996	Spouted fluidized bed	DC plasma torch (4–6 kW); 773–923 K	Ar+H_2+CH_4+O_2 (12 + 3.5 + 0.14 + 0.0035L/min)	Silicon 1.0–1.8 mm; 5 g	PECVD (diamond coating)		Controllable bed quenching condition in	(Liu et al. 1996)

(continued)

Table 9.1 (continued)

Researchers (country)	Year	Operation conditions			Main results		Highlights	Ref.
		System	Plasma source and power input; temperature(K)	Carrier gas; flow rate; pressure;	Materials; particle feature (solid mass)	Application	Main products	
								nanosize particle synthesis
Park and Kim (Korea)	1998	Fluidized bed	RF plasma	Oxygen	HDPE, d_p = 231 μm;	Plasma polymerization		Simple and rapid	(Park and Kim 1998)
Morstein et al. (Switzerland)	2000	Circulating bed	Microwave; 323–393 K;	Argon/oxygen mixture as process gas; 3–5 m/s; 6–20 mbar	Glass beads with diameter of 125 μm	Coating of particles	UltrathinTiO$_2$-film coatings on glass beads	Higher deposition rates	(Morstein et al. 2000)
Sanchez et al. (France)	2001	Spouted bed	Gliding arc; 450 K; 500 W	Nitrogen; 2 sm^3/h	Silica sand or corundum of mean diameter 0.9 mm	PECVD	Coatings of TiN, SiN and SiO$_x$ were deposited 5–25 mm/h	The coating composition may be adjusted from oxides to nitrides	(Sanchez et al. 2001)
Jung et al. (Korea)	2001	Circulating fluidized bed	RF plasma (250 W)	Oxygen+Argon; 0.47 m/s; 133 Pa	HDPE, d_p = 231 μm; 45 g	Surface functionalities		Higher energy efficiency	(Jung et al. 2001)
Leroy et al. (France)	2003	Fluidized bed	Microwave	Nitrogen+oxygen 0.1–2.5 Nl/min	Polyethylene powder 283 μm	Surface modification of polymer powder	Polyethylene powder 312 μm		(Leroy et al. 2003)
Vivien et al. (France)	2002	Fluidized bed	Microwave (300 W)	Nitrogen+Oxygen; 400sccm	PE powder 280 ± 30 μm	Surface modification	Thin film of composition	Allowing the treatment of larger quantities of powder	(Vivien et al. 2002)
Heintze et al. (Germany)	2003	Fluidized bed	RF power supply; (250 W)	Ar–O and Ar–CO$_2$; 0.1–40 mbar	Carbon fibres	Surface functionalisation		Continuous; homogeneous; large amounts of VGCF	(Heintze et al. 2003)
Tatoulian et al. (France)	2004	Fluidized bed	RF source (10–40 W.)	Allylamine/argon gas; P = 40 Pa; 10 cm^3/min	Low density polyethylene (LDPE) Powders 350 um) and non-porous monodisperse polystyrene (PS) beads (200 μm)	Polymerization	Primary amines onto polyethylene and polystyrene; thickness of 400 nm	Amines deposited, amine selectivity, and film insolubility in organic solvents	(Tatoulian et al. 2004)
Jung et al. (Korea)	2004	Circulating fluidized bed	RF plasma 200 W	133 Pa; CF4+He; 0.47 m/s	HDPE, dp = 231 μm; 45 g	Surface functionalities		Higher energy efficiency	(Jung et al. 2004)

(continued)

Table 9.1 (continued)

Researchers (country)	Year	Operation conditions				Main results		Highlights	Ref.
		System	Plasma source and power input; temperature(K)	Carrier gas; flow rate; pressure;	Materials; particle feature (solid mass)	Application	Main products		
Jung et al. (*Korea*)	2004	Circulating Fluidized Bed	RF (150–350 W)	He, Ar, O$_2$, and TEOS; 1200 cm^3/min	Al$_2$O$_3$, 60 μm	PECVD	The layer thickness is approximately 2–4 μm	Stable and uniform	(Jung et al. 2004)
Mutel et al. (France)	2004	Fluidized bed	Microwave plasma (1.5 kW)	N$_2$+O$_2$; 33Nm^3/h	Polyethylene powder with the granulometry (280 μm)	Surface modification to increase its wettability	Oxygen and nitrogen contents on the surface		(Mutel et al. 2004)
Bretagnol et al. (France)	2004	Fluidized bed	RF (20 W)	Nitrogen and ammonia+air; 30 sccm; 50 Pa	Low density polyethylene (LDPE) powders (density: 920 kg m^3, size: 350 lm)				(Bretagnol et al. 2004)
Tap and Willert-Porada (Germany)	2004	Circulating fluidized bed	Microwave plasma 6 kW; 973–1173 K	TiN, TiCl+H$_2$+N$_2$; 30 min	Si–powder; 250–500μm; t–ZrO; 0.2–50 μm; Carbon fibers; length 180 μm, diameter 8 μm	PECVD	Ti coating on different particles	Higher temperatures and lower fluidization velocity values improve the coating density and mechanical stability	(Tap and Willert-Porada 2004)
Arpagaus et al. (Switzerland)	2005	Downer reactor	RF power (300 W); < 363 K	O$_2$/Ar gas mixture 100–1000 sccm; 0.7–10 mbar	A mean particle diameter of 56 lm for HDPE and 47 lm for Co–PA (10 kg/h)	Surface Modification	HDPE and Co–PA powder; Water contact angles of HDPE and Co–PA powder were reduced down to 72 °C and 76 °C	No additional agitation; high homogeneity; short exposure times; semicontinuous or continuous process	(Arpagaus et al. 2005)
Zhu et al. (China)	2005	Fluidized bed	RF power (55 W); monomer	Acrylic acid (AA) and carrier gas (argon)	Nanoscale anatase TiO$_2$ nanoparticles with diameters ranging from 10–30 nm	Plasma polymerization	Deposited AA thin film on the surface of TiO2nanoparticles	Improves the dispersion behavior of TiO$_2$ nanoparticles	(Zhu et al. 2005)
Willert-Porada (Germany)	2006	Fluidized bed	Microwave (3 kW)	(Ar+5 vol.% H$_2$); 200 °C 50, 70 and 90 Nl/min	Short carbon fibres (SGL, Meitingen, Germany; L = 500 μm, φ10 μm) or activated porous	CVD of SG–Silicon		The FBR improves the coating efficiency and has the potential for high throughput	(Willert-Porada 2006)

(continued)

Table 9.1 (continued)

Researchers (country)	Year	Operation conditions				Main results		Highlights	Ref.
		System	Plasma source and power input; temperature(K)	Carrier gas; flow rate; pressure;	Materials; particle feature (solid mass)	Application	Main products		
					carbon powders (Fluka or Merck, Germany, fraction 100–500 µm agglomerate size				
Willert-Porada (Germany)	2006	Fluidized bed	Microwave 3 kW	Ar/H_2+TiCl4; 50, 70 and 90 Nl/min	Short carbon fibres (SGL, Meitingen, Germany; L = 500 µm, φ10 µm)	Coating of carbon materials with TiC		Enhance heterogeneous nucleation of Si on Sili-con seeds on the expense of homogeneous nucleation, leading to an improved process control and yield	(Willert-Porada 2006)
Chen et al. (China)	2006	Fluidized bed	Microwave 11 W	He; 0.05 m/s	Calcium carbonate powders with100 µm in diameter	Polymerization	HMDSO polymer	A relatively uniform and low gas temperature plasma; low energy cost	(Chen et al. 2006)
Rohr and Borer (Switzerland)	2007	Circulating fluidized bed	Microwaves 400 W; 60 °C	Ar+O_2(700+200sccm) 1.5 mbar	Salt NaCl (cubic shape) 207 µm			Simple operation; a plug flow improve the fluidization behavior of particles which tend to agglomerate	(Rohr and Borer 2007)
Li et al. (China)	2008	Fluidized bed	A compact plasma torch 200–500 W	Air 254 °C 35–105°L/h	Glass tube	Synthesis of titania nanoparticles	Titania nanoparticles	Decreasing CO_2 and pollutants release, easily selecting raw materials, and ability to synthesize various nanoparticles	(Li et al. 2008)
Lynda. Aiche1, et al. (France)	2008	Fluidized bed	2.45 GHz microwave 300 W	N_2 267 and 1000 sccm	Polyethylene (PE) powder, 280 µm	Surface modification	Thin layer of SiO_2 on the powders	Continuous	(Alfaro-López HM 2008)

(continued)

Table 9.1 (continued)

Researchers (country)	Year	Operation conditions				Main results		Highlights	Ref.
		System	Plasma source and power input; temperature(K)	Carrier gas; flow rate; pressure;	Materials; particle feature (solid mass)	Application	Main products		
Pajkic and Willert-Porada (Germany)	2008	Fluidized bed	Microwave plasma source; 1273 K		Carbon fibers with saverage diameter d = 7.5 μm and monocrystalline synthetic diamond powder with average particle size d_p = 100 μm; aluminum powder with d_p = 20 μm	Surface PVD	Deposition of thin AlN and Al coatings on particulate carbon materials	Quick and simple production of functional composite particles; simplification and cost reduction of the CVD	(Pajkic and Willert-Porada 2008)
Van de Peppel et al. (Netherlands)	2009	Circulating fluidized bed	DBD	Nitrogen; 43.5 L/min	CuO particles (20–30 micron)	HMDSO coating		Undisturbed fluidized bed regime, low threshold gas flow rate better temperature control	(Van de Peppel et al. 2009)
Pajkic and Willert-Porada (Germany)	2009	Fluidized bed	Microwave; 673–1173 K	Nitrogen; 200–500 Nl h^{-1}	Short carbon fibers Sigrafil C25 M250 UNS, d = 7.5 μm	PECVD(metal coating)	AlN coating more than 95% of its Al-content could be deposited onto the carbon substrates	Continuous production lines; simplifying and/ or eliminating entirely the load-locking procedures as well	(Pajkic and Willert-Porada 2009)
Kim et al. (Korea)	2009	Circulating fluidized bed	RF	Helium gas+Argon+TTIP; 0.55 m/s	Silica gel (SiO$_2$, d_p = 100 μm, Merck) powder	PECVD (metal coating)	Deposition of TiO$_2$films on silica gel powders	Evenly distribution	(Kim et al. 2009)
Chen et al. (China)	2009	Fluidized bed	DBD 7–20 W; 298–302 K	He gas; 0.5 m s^{-1}	Nickel NPs(10–50 nm)	Polymerization nanolayer biofilm	NPs coated with a thicker PAA film (> 8 nm)	Excellent gas fluidity and organic polymerizing characteristics	(Chen et al. 2009)
Laroussi (France)	2009	Fluidized bed	Microwave plasma 300 W; < 363 °C	Nitrogen/oxygen; 10 Torr	High density polyethylene (PE) particles, 260 μm in Sauter diameter and 930 kg/m^3 in grain density	Surface modification	Increases the hydrophilicity of the powders	Low temperature process	(Laroussi 2009)

(continued)

Table 9.1 (continued)

Researchers (country)	Year	Operation conditions				Main results		Highlights	Ref.
		System	Plasma source and power input; temperature(K)	Carrier gas; flow rate; pressure;	Materials; particle feature (solid mass)	Application	Main products		
Deb et al. (USA)	2009	Fluidized bed	ICP discharge 1230–1400 W	Precursor monomers into the cone using a carrier gas, 50sccm	Commercial TiO_2 powders with diameter of 5, 30 and 50 nm (3 g)	Surface functionalization		More stable	(Deb et al. 2009)
Suzuki et al. (Japan)	2009	Fluidized bed	DC arc plasma (180 W); 2273 K	Nitrogen; 0.2–0.5m3/h	TiO_2 (5 nm) BET = 200–220 m^2/g	Surface modification of nanopowder	Nitrogen-doped TiO_2 nanopowder	Low cost operation	(Suzuki et al. 2009)
Song et al. (Korea)	2011	Fluidized bed	RF power (100 W); 333 K	Argon 0.5 torr; 5 min 0.216mn/s	Polystyrene powder d_p = 237 µm(160 g)			Synthesis of core-shell structure of PEG–g– PS powder	(Song et al. 2011)
Jafari et al. (France)	2011	Fluidized bed	RF power (10–27 W)	Mixture of acrylic acid/argon (6 + 18 sccm); 50 Pa	Spherical monodisperse polystyrene beads of 207 µm in diameter	Plasma polymerization		Plasma polymerized acrylic acid (PPAA) coatings on the surface of polystyrene beads	(Jafari et al. 2011)
Roth et al. (Switzerland)	2012	Fluidized bed	ICP source (200–300 W)	HMDSO+Ar+O_2; 50 + 500 + 950 sccm	Polycarbon-ate with admixed graphite	Surface modification	Coated with an approximately 2 nm thick platinum layer	A very short process time in the order of 1 s	(Roth et al. 2012
Von Ommen et al. (France)	2011	Circulating fluidized bed	DBD (10–300 W)	Nitrogen	20–30 µm CuO particles	PECVD	20–30 µm CuO particles can be provided with a thin SiOxfilm	Temperature-sensitive powder materials can be treated and obtained	(Von Ommen et al. 2010)
Jafari et al. (France)	2011	Fluidized bed	Inductively coupled glow discharge; 50 Pa; Q_{Ar} = 18 sccm		Spherical monodisperse polystyrene bead LCC Engineering & Trading; GmbH) of 207 lm in diameter	Polymerization	Plasma polymerized acrylic acid coating		(Jafari et al. 2011)
Soto et al. (Mexico)	2011	Fluidized bed	Microwave plasma(200 W)	80–666.6 Pa; Argon; 100 sccm; 10 min; 423 K	90% silica–10% alumina	Synthesis of silver nanoparticles	Grow stable nanoparticles with an average 13 nm diameter	The inherent simplicity	(Soto et al. 2011)

(continued)

Table 9.1 (continued)

Researchers (country)	Year	Operation conditions				Main results		Highlights	Ref.
		System	Plasma source and power input; temperature(K)	Carrier gas; flow rate; pressure;	Materials; particle feature (solid mass)	Application	Main products		
Bartolomeu et al. (France)	2011	Fluidized bed	Microwave; 473 K	Argon	Powders (355–500 μm)				(Bartolomeu et al. 2011)
Abadjieva et al. (Netherland)	2012	Circulating Fluidized Bed	VDBD; 2.5 m/s; atm; argon 50 L/min+precursor		Glass beads had a diameter in the range between 40 and 70 μm	Polymers	Coating of the glass beads with a fluorocarbon containing layer	Operate under atmospheric pressure	(Abadjieva et al. 2012)
Caquineau et al. (France)	2012	Fluidized bed	Microwave plasma (300 W); 373 K	Nitrogen and oxygen gasses +Silane; 800sccm; 10 Torr	High density polyethylene (PE) powders 260 μm; 50–100 g	PECVD		Temperature-sensitive polyethylene (PE) powders were successfully coated by silicon oxide	(Caquineau et al. 2012)
Put et al. (Belgium)	2013	Downer bed	DBD plasma	Nitrogen+air (about 3 slm); 0.1 g powder; 25 ms	Ultra high molecular weight polyethylene powder particles 150 μm (20–42.5 g)	Surface modification	Oxygen and nitrogen contents on the surface	Stable	(Put et al. 2013)

References

Abadjieva E, van der Heijden AEDM, Creyghton YLM, van Ommen JR. Fluorocarbon coatings deposited on micron-sized particles by atmospheric PECVD. Plasma Process Polym. 2012;9 (2):217–24.

Arnauld P, Cavadias S, Amouroux J. The interaction of a fluidized bed with a thermal plasma: application to limestone decomposition. In: 7th international symposium on plasma chemistry (Eindhoven, 1985). 1985.

Arpagaus C, Sonnenfeld A, von Rohr PR. A downer reactor for short-time plasma surface modification of polymer powders. Chem Eng Technol. 2005;28(1):87–94.

Bai XY, Yi WM, Wang LH, Li YJ, Cai HZ. Fast pyrolysis of corn stalk for bio-oil in a plasma heated fluidized bed. Trans CSAE. 2005;21(12):127–30.

Bao WR, Chang LP, Lu YK. Study on main factors influencing acetylene formation during coal pyrolysis in arc plasma. Process Saf Environ. 2006;84(3):222–6.

Bartolomeu R, Foix M, Fernandes A, Tatoulian M, Ribeiro MF, Henriques C, et al. Fluidized bed plasma for pre-treatment of co-ferrierite catalysts: an approach to NO_x abatement. Catal Today. 2011;176(1):234–8.

Bretagnol F, Tatoulian M, Arefi-Khonsari F, Lorang G, Amouroux J. Surface modification of polyethylene powder by nitrogen and ammonia low pressure plasma in a fluidized bed reactor. React Funct Polym. 2004;61(2):221–32.

Butscher D, Schlup T, Roth C, Müller-Fischer N, Gantenbein-Demarchi C, von Rohr PR. Inactivation of microorganisms on granular materials: reduction of Bacillus amyloliquefaciens endospores on wheat grains in a low pressure plasma circulating fluidized bed reactor. J Food Eng. 2015;159:48–56.

Caquineau H, Aiche L, Vergnes H, Despax B, Caussat B. Low temperature silicon oxide deposition on polymer powders in a fluidized bed coupled to a cold remote plasma. Surf Coat Tech. 2012;206(23):4814–21.

Chen G, Chen S, Zhou M, Feng W, Gu W, Yang S. Application of a novel atmospheric pressure plasma fluidized bed in the powder surface modification. J Phys D Appl Phys. 2006;39 (24):5211.

Chen G, Zhou M, Chen S, Lv G, Yao J. Nanolayer biofilm coated on magnetic nanoparticles by using a dielectric barrier discharge glow plasma fluidized bed for immobilizing an antimicrobial peptide. Nanotech. 2009;20(46):465706.

Currier R, Blacic J, Trkula M. Hydrogen plasma reduction of planetary materials. In: ISRU III technical interchange meeting (Denver, 1999). 1999.

Currier R, Blacic J. Plasma processing of lunar and planetary materials. In: Space resources roundtable II (Colorado, 2000). 2000.

Deb B, Kumar V, Druffel TL, Sunkara MK. Functionalizing titania nanoparticle surfaces in a fluidized bed plasma reactor. Nanotech. 2009;20(46):465701.

El-Naas MH, Munz RJ, Ajersch F. Modelling of a plasma reactor for the synthesis of calcium carbide. CAN Metall Quart. 1998;37(1):67–74.

El-Naas MH. Synthesis of calcium carbide in a plasma spout fluid bed. McGill University; 1996.

Emome A, Jurewize T. Fuel synthesis for solid oxide fuel cells by plasma spouted bed gasification. In: 14th international symposium on plasma chemistry (Prague, 1999). 1999.

Heintze M, Brüser V, Brandl W, Marginean G, Bubert H, Haiber S. Surface functionalisation of carbon nano-fibres in fluidised bed plasma. Surf Coat Tech. 2003;174–175(3):831–4.

Jafari R, Tatoulian M, Arefi-Khonsari F. Improvement of the stability of plasma polymerized acrylic acid coating deposited on PS beads in a fluidized bed reactor. React Funct Polym. 2011;71(4):520–4.

Jung SH, Park SH, Dong HL, Sang DK. Surface modification of HDPE powders by oxygen plasma in a circulating fluidized bed reactor. Polym Bull. 2001;47(2):199–205.

Jung SH, Park SH, Kim SD. Surface treatment of polymeric fine powders by CF_4 plasma in a circulating fluidized bed reactor. J Chem Eng Jpn. 2004a;37(2):166–73.

Jung SH, Sang MP, Park SH, Sang DK. Surface modification of fine powders by atmospheric pressure plasma in a circulating fluidized bed reactor. Ind Eng Chem Res. 2004b;43(18):5483–8.

Karches M, Bayer C, von Rohr PR. A circulating fluidized bed for plasma-enhanced chem vapor depos on powders at low temperatures. Surf Coat Tech. 1999;116–119(4):879–85.

Kim GH, Kim SD, Park SH. Plasma enhanced chem vapor depos of TiO_2 films on silica gel powders at atmospheric pressure in a circulating fluidized bed reactor. Chem Eng Process. 2009;48(6):1135–9.

Kroker T, Kolb T, Schenk A, Krawczyk K, Młotek M, Gericke KH. Catalytic conversion of simulated biogas mixtures to synthesis gas in a fluidized bed reactor supported by a DBD. Plasma Chem Plasma P. 2012;32(3):565–82.

Laroussi M. Low-temperature plasmas for medicine? IEEE T Plasma Sci. 2009;37(6):714–25.

Lee H, Sekiguchi H. Plasma-catalytic hybrid system using spouted bed with a gliding arc discharge: CH_4 reforming as a model reaction. J Phys D Appl Phys. 2011;44(27):274008.

Leroy JB, Fatah N, Mutel B, Grimblot J. Treatment of a polyethylene powder using a remote nitrogen plasma reactor coupled with a fluidized bed: influence on wettability and flowability. Plasmas Polym. 2003;8(1):13–29.

Li MW, Gonzalez-Aguilar J, Fulcheri L. Synthesis of titania nanoparticles using a compact nonequilibrium plasma torch. Jpn J Appl Phys. 2008;47(9):7343–5.

Li X, Han J, Wu CN, Guo Y, Yan B, Cheng Y. Coal tar pyrolysis to acetylene in thermal plasma. CIESC J. 2014;65(9):3680–6.

Liang M, Chen J, Lin WM, Liu K. Thermodynamics study on solid phase decarbonization of high carbon ferromanganese powder. Ferro-Alloys. 2009.

Liu LX, Rudolph V, Litster J. A direct current, plasma fluidized bed reactor: its characteristics and application in diamond synthesis. Powder Technol. 1996;88(1):65–70.

Mersereau OS. Plasma spout-fluid bed calcination of lac dore vanadium ore concentrate. McGill University; 1990.

Młotek M, Sentek J, Krawczyk K, Schmidt-Szałowski K. The hybrid plasma-catalytic process for non-oxidative methane coupling to ethylene and ethane. Appl Catal A-Gen. 2009;366(2): 232–41.

Mohammedi MN, Cavvadias S, Leuenberger JL, Francke E, Amouroux J. CO-processing conception of a quenching plasma reactor associated to a fluidized bed designed to hydrogen radical production at low temperatures. Plasma Chem Plasma P. 1995;16(1):191–210.

Morstein M, Karches M, Bayer C, Casanova D, von Rohr PR. Plasma CVD of ultrathin TiO_2 films on powders in a circulating fluidized bed. Chem Vapor Depos. 2000;6(1):16–20.

Mutel B, Bigan M, Vezin H. Remote nitrogen plasma treatment of a polyethylene powder: optimization of the process by composite experimental designs. Appl Surf Sci. 2004;239 (1):25–35.

Pajkic Z, Willert-Porada M. Arc-PVD coating of powders in a microwave plasma fluidized bed. In: IEEE 35th international conference on plasma science (Karlsruhe, 2008). 2008.

Pajkic Z, Willert-Porada M. Atmospheric pressure microwave plasma fluidized bed CVD of AlN coatings. Surf Coat Tech. 2009;203(20):3168–72.

Park SH, Kim SD. Oxygen plasma surface treatment of polymer powder in a fluidized bed reactor. Colloid Surface A. 1998;133(1–2):33–9.

Park SH, Kim SD. Functionalization of HDPE powder by CF_4 plasma surface treatment in a fluidized bed reactor. Korean J Chem Eng. 1999;16(6):731–6.

Put S, Bertels C, Vanhulsel A. Atmospheric pressure plasma treatment of polymeric powders. Surf Coat Tech. 2013;234(10):76–81.

Ramachandran K, Kikukawa N. Plasma in-flight treatment of electroplating sludge. Vacuum. 2000;59(1):244–51.

Roth C, Keller L, von Rohr PR. Adjusting dissolution time and flowability of salicylic acid powder in a two stage plasma process. Surf Coat Tech. 2012;206(19–20):3832–8.

Sanchez I, Flamant G, Gauthier D, Flamand R, Badie JM, Mazza G. Plasma-enhanced chem vapor depos of nitrides on fluidized particles. Powder Technol. 2001;120(1–2):134–40.

Schmidt-Szałowski K, Krawczyk K, Młotek M. Catalytic effects of metals on the conversion of methane in gliding discharges. Plasma Process Polym. 2007;4(7–8):728–36.

Sekiguchi H, Matsudera N, Kanzawa A. Destruction of CH_3Cl using plasma fluidized CaO bed. In: 12th international symposium on plasma chemistry (Minnesota, 1995). 1995.

Song LH, Park SH, Jung SH, Sang DK, Park SB. Synthesis of polyethylene glycol-polystyrene core-shell structure particles in a plasma-fluidized bed reactor. Korean J Chem Eng. 2011;28 (2):627–32.

Soto G, Tiznado H, Contreras O, Pérez-Tijerina E, Cruz-Reyes J, Valle MD, et al. Preparation of a Ag/SiO_2 nanocomposite using a fluidized bed microwave plasma reactor, and its hydrodesulphurization and Escherichia coli bactericidal activities. Powder Technol. 2011;213(1–3):55–62.

Steinbach PB, Manahan SE, Larsen DW. The chemical reduction of small inorganic gases in an electrothermal plasma reactor. Microchem J. 2003;75(3):223–31.

Suzuki Y, Gonzalez-Aguilar J, Traisnel N, Berger MH, Repoux M, Fulcheri L. Non-equilibrium nitrogen DC-arc plasma treatment of TiO_2 nanopowder. J Nanosc Nanotechnol. 2009;9 (1):256–60.

Takarada T, Tamura K, Takezawa H, Nakagawa N, Kato K. The effect of pretreatment in a fluidized bed upon diamond synthesis on particles by chemical vapour deposition. J Mater Sci. 1993;28(6):1545–50.

Tap R, Willert-Porada M. Dual PE-CVD circulating fluidized bed reactor. IEEE T Plasma Sci. 2004;32(5):2085–92.

Tatoulian M, Brétagnol F, Arefi-Khonsari F, Amouroux J, Bouloussa O, Rondelez F, et al. Plasma deposition of allylamine on polymer powders in a fluidized bed reactor. Plasma Process Polym. 2004;2(1):38–44.

Tsukada M, Goto K, Yamamoto RH, Horio M. Metal powder granulation in a plasma-spouted/ fluidized bed. Powder Technol. 1995;82(3):347–53.

Vaidyanathan A, Mulholland J, Ryu J, Smith MS, Circeo LJ. Characterization of fuel gas products from the treatment of solid waste streams with a plasma arc torch. J Environ Manage. 2007;82 (1):77–82.

van de Peppel R, Abadjieva E, van der Heijden AEDM, Creyghton Y. Plasma coating process for powder surface modification. In: 19th international symposium on plasma chemistry (Bochum, 2009). 2009.

Vivien C, Wartelle C, Mutel B, Grimblot J. Surface property modification of a polyethylene powder by coupling fluidized bed and far cold remote nitrogen plasma technologies. Surf Interface Anal. 2002;34(1):575–9.

von Ommen JR, Abadjieva E, Creyghton YL. Plasma-enhanced chemical vapour deposition on particles in an atmospheric circulating fluidized bed. In: 13th international conference on fluidization-new paradigm in fluidization engineering (Korea, 2010). 2010.

von Rohr PR, Borer B. Plasma-enhanced CVD for particle synthesis using circulating fluidized bed technology. Chem Vapor Depos. 2007;13(9):499–506.

Wang LH, Bai XY, Yi YM, Kong FX, Li YJ, He F, Li ZH. Characteristics of bio-oil from plasma heated fluidized bed pyrolysis of corn stalk. Trans CSAE. 2006;663–665(3):502–5.

Wang Q, Cheng Y, Jin Y. Dry reforming of methane in an atmospheric pressure plasma fluidized bed with $Ni/\gamma-Al_2O_3$ catalyst. Catal Today. 2009;148(3):275–82.

Willert-Porada M. Advances in microwave and radio frequency. Berlin: Springer; 2006.

Yan B, Cheng Y, Jin Y. Cross-scale modeling and simulation of coal pyrolysis to acetylene in hydrogen plasma reactors. AIChE J. 2013;59(6):2119–33.

Yan B, Xu P, Guo CY, Jin Y, Cheng Y. Experimental study on coal pyrolysis to acetylene in thermal plasma reactors. Chem Eng J. 2012;207–208(10):109–16.

Zhang CM, Liu RH, Yi WM, Sun Q. Experiment on plasma pyrolysis of corn stalk for liquid fuel. Trans CS AM. 2009.

Zhu CW, Zhao GY, Hlavacek V. A d.c. plasma-fluidized bed reactor for the production of calcium carbide. J Mater Sci. 1995;30(9):2412–19.

Zhu F, Zhang J, Yang Z, Guo Y, Li H, Zhang Y. The dispersion study of TiO_2 nanoparticles surface modified through plasma polymerization. Physica E. 2005;27(4):457–61.

Chapter 10
Applicative Ability and Environmental Risk

Abstract This chapter analyses the application and environmental risks of plasma fluidized bed. Firstly, energy analysis and economic analysis are mentioned. Plasma fluidized bed has more advantages than traditional processes. Secondly, the application ability and environmental risk are mentioned. The problem of plasma fluidized bed has not been reported in the relevant literature. Therefore, further research and discovery are needed in this area. Finally, the prospect of future application of plasma fluidized bed is presented in this chapter, and several existing problems are proposed, which need further development and improvement.

Keywords Energy analysis · Economic analysis · Technical risk
Environmental risk · Outlook

10.1 Energy and Economy

10.1.1 Energy Analysis

Energy needed for reactions within plasma fluidized bed fundamentally is generally supplied by plasma source. For operation in thermal plasma fluidized bed, thermal energy can be efficiently offered to heat the particles to specific temperature. Also, active species generated during plasma discharge can effectively enhance the activity of the materials and lower the temperature caller for expected reaction. Moreover, uniform transfer in the plasma fluidized bed ensures the fully utilization of energy. Therefore, the energy efficiency of thermal plasma fluidized bed is much higher than conversional thermal process or plasma furnace. Taking a DC torch plasma used for the synthesis of calcium carbide, energy efficiency is 40% higher than plasma furnace and the operation temperature is much lower than that of the conventional thermal operation. Similarly, for nonthermal plasma fluidized bed, it has been pointed out that the energy efficiency compared with other chemical technology. However, the energy cost by fluidization of particles is another important aspect, which has not been investigated in quantitatively.

© Springer Nature Singapore Pte Ltd. and Zhejiang University Press 2018

C. Du et al., *Plasma Fluidized Bed*, Advanced Topics in Science and Technology in China, https://doi.org/10.1007/978-981-10-5819-6_10

10.1.2 Economic Analysis

As a high energy efficiency reactor, the construction of plasma fluidized bed is very simple and easy to attain, the material needed to manufacturing plasma fluidized bed is also easy to get, especially for low temperature reactor. Also, the energy efficiency of plasma fluidized bed is relatively higher than other comparable process. For example, for treatment of advanced materials, the products are generally of high value and have economic advantages and operation cost is much lower than other chemical process such as techniques due to the lack of chemicals import such organic solvent. Of course, for some low pressure plasma fluidized bed such as some RF or microwave fluidized bed, the energy cost of the vacuum system is another important problem. However, generally, plasma fluidized bed is of great energy potential for further application, particularly for those with high value products.

10.2 Applicative Ability and Environmental Risk

10.2.1 Technical Risk

It is not so certain what further possible issues are involved since hardly any problems associated with the reactor or during continuous synthesis are reported. However, it can be sure in each and every research, and there will be problems that are omitted or which most authors fail to identify. Commonly affected problems are the mechanics of the particulate bed and the flow ability of the powders such as surface activation, solidifying bridge between particles, internal rupture of particles and caking. These are attributed to particle size, shape, environmental conditions, cohesiveness, surface roughness, electromagnetic force, orifice diameter, wall effects and the angle of inclination, etc. For arc discharge synthesis, discussion of the problem associated with particle-reactor interaction is hardly to be found and possibly there is none at all. For the plasma arc, it is possible that some of the studies have a certain level of similarity to CVD, but only in terms of continuous synthesis during the CNT discharge.

One other issue which could be considered as important, other than the process efficiency described earlier, which has been neglected by most researchers, is the study of how the particles and discharge interact within the reactor. To our knowledge so far this is studied limitedly. Therefore, the knowledge discharge characteristics, the hydrodynamic as well as the heat transfer and mass transfer of plasma fluidized bed are actually very limited and unable to the be suitable for various kinds of plasma fluidized beds, which of key role to instruct the scale up and application of such a reactor. And also, the melting and aggregating is always a problem needed to deal with for ultrahigh temperature process.

One major consideration is the ability of the latter to be fluidized. Although for some specific applications, such as production of catalytic materials, the powders belong to Geldart's class A or B categories, and can therefore be easily fluidized. Most powders of interest in advanced materials usually fall into Geldart's class C category. These powders are very cohesive and difficult to handle, to fluidize and to process in a non-agglomerated form. High aspect ratios handling and sometimes fluidization, and most fine powders are frequently handled, coated or modified in an agglomerated form. Thus, to efficiently design a FB-CVD process, the constraints linked to fluidization have to be taken into account. Spouted beds, circulating beds operating in turbulent and fast-transport regimes or vibro-fluidized beds are possible alternatives to classical fluidized-bed reactors.

The lifetime of the reactor, including the metal electrodes, the reactor chamber and the gas distributors may be another obstacle for the development of plasma fluidized bed because plasma fluidized bed is always immersed in highly reactive chemical atmosphere or the high temperature environment.

10.2.2 *Environmental Risk*

For plasma fluidized bed, the exhausted gas handling is common for the process in which many problems encountered are attributed to the characteristic of reaction between injected gas and the particles or unreacted process gas. To illustrate, for PECVD process with plasma fluidized bed, the used precursor sometimes is toxic, such as HMVCD, and most of time the process gas can be utilized totally before discharging, the treatment of the exhausted gas is of importance.

Also, for high temperature process inside plasma fluidized bed, nonuniform heat transfer in local place is a problem needed to overcome, for hot zone is usually easy to form, which can result in serious accident such as explosion.

10.3 Outlook

The potential of the plasma fluidized bed technology is high since it covers a wide range of operations, including: (1) metallurgy extraction, (2) green energy production (3) environmental protection and (4) advanced material. It can be reasonably expected that, in the near future plasma fluidized bed technology will contribute to intensifying the industrial development of powder treatment.

Nowadays, investigations of plasma fluidized bed are focus on the various applications, mechanism such as hydrodynamic, discharge mechanism, or heat transfer and mass transfer. France scientists have also devoted a lot for the first proposal the plasma fluidized bed and then open up the new era of development of plasma fluidized bed. Switzerland scientists devoted a lot with respect of the application in advanced materials, especially in the field of surface modification.

France scientists and Chinese scientists have made up the field of hydrodynamic and transfer mechanism of plasma fluidized bed to offer further insight of plasma fluidized bed. Also Japan and Korea scientists also studied deeply into the methane reformation by plasma fluidized bed. With deepened development of plasma fluidized bed, noval reactors have become the huge demand for a complex reaction processes into some specific processes, which call for a responsible use of such a reactor which designated for the specific applications (metallurgy and material etc.) is needed. A significant amount of basic research has to be done now and in future to identify potential indications and to estimate the risks of plasma fluidized bed used in order to apply a plasma fluidized bed reactor stably and efficiently, which includes the proposal and investigation of proper reactor for specific applications, scale-up methodology, multiphase modeling and discharge modeling for general plasma fluidized bed and simulation incorporating with reacting flows, measurement techniques to capture detail flow and discharge phenomenon and fast reaction process in kinetics.

A necessary next step is to make such a set of basic hydrodynamic characteristic as well as chemical reactions parameters mandatory and to transfer it into legal rules and standards including a risk analysis considering different field of potential applications.

However, besides all technical as well as hydrodynamic and chemical details and features, decision if a plasma fluidized bed may be useful for the application should be made using the following 10 questions:

(1) Is it a high efficiency and energy efficiency in industrial process?
(2) Is the absence of undesirable local or systemic side effects (explosion or other accidents) proved?
(3) Can the reactor work stably and safety in a long span?
(4) Is the product selective and product quality controllable?
(5) Is the discharge process, the hydrodynamic and the transfer between the plasma gas and the injected particles as well as the chemical reaction predictable for the instruction of the production?
(6) Is the conversion of the raw materials acceptable?
(7) Is it able to deal with the exhausted gas of the system easily for environmental protection?
(8) Is it economical process?
(9) Is the scale of the reactor big enough to meet the demand of the manufacture?
10 Are there no simpler and economical alternative process to the same application?

Only if the majority of these questions can be affirmed, a further development of plasma fluidized bed for industrial application is possible. The key for this aim lies in both a comprehensive and careful discharge and hydrodynamic as well as heat and mass transfer characterization and optimization of the plasma fluidized bed reactors.

Plasma fluidized bed is a new field with enormous opportunities for significant research potential and industrial application potential. It is broad enough to promote cooperation and is an opportunity.

The manufacturer's authorised representative in the EU is Springer
Nature Customer Service Centre GmbH, Europaplatz 3, 69115 Heidelberg,
Germany. If you have any concerns regarding our products, please
contact ProductSafety@springernature.com

Printed and bound by CPI Group (UK) Ltd, Croydon, CR0 4YY

28/11/2025

02007725-0006